Lecture Notes in Economics and Mathematical Systems

534

Founding Editors:

M. Beckmann
H. P. Künzi

Managing Editors:

Prof. Dr. G. Fandel
Fachbereich Wirtschaftswissenschaften
Fernuniversität Hagen
Feithstr. 140/AVZ II, 58084 Hagen, Germany

Prof. Dr. W. Trockel
Institut für Mathematische Wirtschaftsforschung (IMW)
Universität Bielefeld
Universitätsstr. 25, 33615 Bielefeld, Germany

Editorial Board:

A. Basile, A. Drexl, W. Güth, K. Inderfurth, W. Kürsten, U. Schittko

T0239491

Springer
Berlin
Heidelberg
New York
Hong Kong
London
Milan
Paris
Tokyo

Marco Runkel

Environmental and Resource Policy for Consumer Durables

 Springer

Author

Dr. Marco Runkel
Department of Economics
University of Munich
Ludwigstraße 28/Vgb./III
80539 Munich
Germany

Cataloging-in-Publication Data applied for
Bibliographic information published by Die Deutsche Bibliothek.
Die Deutsche Bibliothek lists this publication in the Deutsche Nationalbibliographie; detailed
bibliographic data is available in the Internet at <http://dnb.ddb.de>.

ISSN 0075-8442
ISBN 3-540-20569-1 Springer-Verlag Berlin Heidelberg New York

Springer-Verlag is a part of Springer Science+Business Media

springeronline.com

© Springer-Verlag Berlin Heidelberg 2004
Printed in Germany

Typesetting: Camera ready by author
Cover design: *Erich Kirchner*, Heidelberg

Printed on acid-free paper   55/3143/du   5 4 3 2 1 0

To Nicole

Preface

For many years, the economic analysis of environmental problems proceeds on the assumption that the quantity of consumption goods is the main variable determining the extent of pollution caused by these goods. Only recently, some economists took into account that certain product characteristics like biodegradability, recyclability or durability play an important role in designing an efficient environmental policy. To get a better understanding of this topic, the research project 'Economic Analysis of Product-Related Environmental Policy' was established in June 1998 at the University of Siegen, Germany. The present book was written during my time as collaborator in this research project. It is identical to my doctoral thesis which the Department of Economics at the University of Siegen accepted in June 2002.

I am extremely grateful to Rüdiger Pethig who was the supervisor of the above research project and my doctoral thesis. He introduced me to this exciting research field and steadily supported my work with a host of excellent ideas and suggestions. By his comprehensive guidance and advice, he created a stimulating atmosphere in which scientific research is a pleasure. I am also indebted to my second supervisor Karl-Josef Koch. He was always open for discussion and gave useful comments which improved my work. My sincere thanks are due to Bernd Huber from the University of Munich for allowing me to finish the doctoral thesis during the first months of my time at his chair and to Walter Buhr who kindly agreed to act as the chairman of my dissertation committee.

The present study has also benefited from the help of my colleagues. The first and foremost thanks go to Thomas Eichner, my colleague in the above research project. He not only provided patient support in fixing ideas, but additionally wrote joint papers with me and became a very good friend. I received helpful comments also from Andreas Wagener, Thomas Steger, Sao-Wen Cheng, Heike Johannes and Daniel Weinbrenner. Furthermore, the financial support of the *Deutsche Forschungsgemeinschaft DFG* (German Research Foundation) is gratefully acknowledged.

My special thanks go to my family, in particular my parents who gave me, at all times, the freedom to do those things I am really interested in. The most important person, however, is my wife Nicole. Although she had to invest a lot of time in her own education, she always supported me and gave me the motivation and power to work on my doctoral thesis. I thank her for her love and her unwavering confidence in me.

Munich,   *Marco Runkel*
October 2003

Contents

1

Introduction

In the last decades, environmental problems have received increasing attention. Eminent examples are global warming, pollution of the world oceans and deforesting of the tropical rain forest. From an economic point of view, environmental problems are externality problems leading to market failure. In order to illustrate this, consider a producer who manufactures a consumption good. On the one hand, this economic activity causes *production costs* as, for example, costs of material, labor and capital inputs. On the other hand, producing the consumption good usually generates emissions like smoke, wastewater or solid waste. The emissions cause environmental damage which leads to *environmental costs*. A market economy typically comprises markets for material, labor and capital. Hence, the producer has to pay for these production inputs and accounts for marginal production costs in her manufacturing decision. In contrast, the market economy usually fails to provide price signals for environmental assets. The producer pollutes the environment for free and does not account for marginal environmental costs in her production decision. Therefore, the price of the consumption good is inefficiently low since it does not reflect full social costs. The inefficiently low price leads to an inefficiently high quantity of the good with the consequence that the production process of the good is too environmental intensive. This argumentation can be found, for example, in Baumol and Oates (1988) or Tietenberg (1996).

Market failure is typically considered a justification of governmental regulation of the economic activity which generates emissions. The first regulatory approach in economic literature was Pigouvian taxation considered by the seminal work of Pigou (1938). The idea of the Pigouvian approach is to tax the emissions generating activity by its marginal environmental costs. This internalizes the environmental externality, i.e. the tax induces producers to account not only for marginal production costs, but also for marginal environmental costs. The consumption good becomes more expensive such that fewer units are produced and sold and such that the amount of emitted pollutants is reduced. In sum, Pigouvian taxation renders the production of the consumption good efficient.

Of course, the economic analysis of environmental policy is more comprehensive. Economic literature additionally inquires into the efficiency properties of other regulatory instruments like e.g. abatement subsidies, tradable emissions permits or deposit-refund systems. A survey is offered by Xepapadeas (1997). Furthermore, economic literature has made great effort to relax the basic assumptions underlying the analysis of Pigouvian taxation. For example, Buchanan (1969) was the first author who considers emissions taxes under conditions of imperfect competition, Weitzman (1974) analyzes the implications of uncertainty for environmental policy and Ko et al. (1992) investigate regulation of a pollutant stock instead of a pollutant flow. A comprehensive survey can again be found in Xepapadeas (1997).

A recent extension of the basic economic analysis of environmental policy is the so-called product-related environmental policy. The rationale of this kind of policy is that the amount of emissions and the associated environmental costs are not only influenced by the quantity of the good produced, but also by particular product characteristics. For example, Cremer and Thisse (1999) investigate a model of vertical product differentiation and interpret the product characteristic considered as the product's environmental quality which influences the environmental damage of the product. Fullerton and Wu (1998) and Choe and Fraser (1999, 2001) model the recyclability of consumption goods explicitly by introducing a product design variable which increases production costs, on the one hand, but raises productivity in the recycling sector, on the other hand. Recycling characteristics of consumption goods are also considered by Eichner and Pethig (1999, 2000, 2001) who focus on the material content of consumption goods and by Calcott and Walls (1999, 2000) who model the product's weight. The main focus of these product-related studies is to identify and characterize market failure when markets for environmental assets, consumption residuals or product characteristics are absent and to determine regulatory schemes capable of restoring efficiency.

The present study aims at contributing to the economic analysis of product-related environmental policy by focusing on environmental aspects of a further product characteristic. Fullerton and Wu (1998, p. 131) and Eichner and Pethig (2001, p. 113) argue that their results are valid not only for non-durable goods (like e.g. foods), but also for consumer durables (like e.g. cars). Nevertheless, their models do not contain product design variables like product lifetime or, more generally, product durability which usually constitute durable consumption goods. This is the starting point of the present study. We introduce product durability as an explicit choice variable of producers and take into account that variations in product durability affect emissions from producing and consuming the durable good. The empirical relevance of such a framework is documented, for example, by Asch and Seneca (1976) who argue that the most heavily polluting U.S. manufacturing sectors are durable good industries, by Hamilton and Macauley (1999) who find significant variations in product durability of U.S. automobiles over the last decades and by the study of OECD (1982, pp. 56) which estimates that constructing a long living 'middle-class' car with a lifetime of 18–25 years reduces material inputs about 55–65% and energy inputs about 5–10%.

To the extent that producers do not take into account the relation between product durability and emissions, they ignore the environmental costs of their durability choice and set product durability inefficiently. The main objective of the present study is to formally identify and characterize such market failure and to determine regulatory policies capable of restoring efficiency. Before outlining the analysis in more detail, the next paragraph gives a brief survey of previous product durability studies.[1]

Previous Economic Literature on Product Durability

The industrial organization literature provides a large number of studies inquiring into the relationship between product durability and market structure. The starting point of this literature is the hypothesis of planned obsolescence. This hypothesis says that a monopolist or an oligopolist takes advantage of her market power not only to restrict the output of the durable good, but also to promote future sales by shortening the lifetime of her products. Focusing on the long run stationary state of a durable good economy, Martin (1962), Kleiman and Ophir (1966), Levhari and Srinivasan (1969) and Schmalensee (1970) claim to have found theoretical arguments in support of the planned obsolescence hypothesis. But they completely ignore the role of the adjustment path in the long run decision of durable good producers. In contrast, the long run analysis of Swan (1970) and Sieper and Swan (1973) takes into account the adjustment path. These authors show that, under certain conditions, product durability in the stationary state is independent of market structure, i.e. the monopolist chooses the same long run product durability as producers under perfect competition. Goering (1992) generalizes this independence result to the case of oligopolistic producers.

Swan (1970) offers the following intuition for his independence result. Suppose producers rent the durable good to consumers. If consumers demand the services of the durable good, then the rental price at a particular point in time equals the consumers' marginal willingness-to-pay for the services at this time. The present value of rental revenues is independent of product durability and, thus, producers choose product durability in a cost-minimizing way. If the production technology is linear in output, then unit costs are represented by a function of durability only, and it can be shown that cost-minimizing product durability does not depend on the output of the durable good. Consequently, cost-minimizing product durability is independent of the producers' market shares and, hence, independent of market structure. Swan (1970) claims that the same rationale holds in case producers sell the durable good rather than rent it since the sales price of the durable equals the present value of rentals

---

[1] We focus only on those studies which are relevant for the subsequent analysis. A comprehensive survey of early articles is given in Schmalensee (1979). More recent articles not required for our analysis are, for example, Waldman (1993, 1996), Hendel and Lizzeri (1999) and Fishman and Rob (2000).

which a unit of the durable generates over the whole lifetime and, thus, profits from selling the durable equal profits from renting the durable.

A large part of the durability literature since the mid 70s inquires into the question whether the independence result remains true if the assumptions of Swan's original analysis are relaxed. Of special interest is the assumption of a linear production technology.[2] Schmalensee (1979) argues that in case of output-dependent unit costs, the independence result remains valid in the long run, but may not hold in the short run (i.e. on the adjustment path to the stationary state). For the short run analysis, he refers to the analysis of Sieper and Swan (1973) who show that fixed capacity costs cause output-dependent unit costs and so make short run product durability conditional on market structure, in general. This result is specified by Muller and Peles (1990): A durable good monopolist chooses an inefficiently *low* product durability in the short run if the stock of the durable good monotonically increases over time. Concerning the long run analysis, Schmalensee (1979) refers to Auernheimer and Savings (1977) who account for adjustment costs which make unit costs dependent on output in the short run, but not in the long run.

More recent studies reexamine Schmalensee's conclusion that the independence result remains valid in the stationary state. Abel (1983) presents a durable good model with a general production technology allowing for both linear and nonlinear cost functions. He derives a sufficient condition for the independence result to hold in the long run. The condition is called Abel condition and says that the ratio of the elasticity of production costs with respect to output and the elasticity of production costs with respect to product durability is independent of output. Under this condition, cost-minimizing product durability is not influenced by the market share of producers and, thus, is independent of market structure. The Abel condition is satisfied e.g. by all production technologies which are linear in output and, more generally, by all Cobb-Douglas production technologies. Nevertheless, it is possible to identify production technologies which violate the Abel condition and which upset the independence result. Such a case is presented by Goering (1993). He takes into account cost reductions through learning effects in the production of the durable good. Owing to the learning effects, unit production costs depend on accumulated output so that cost-minimizing product durability is influenced by the market shares of producers. In sum, we conclude that in the short run as well as in the long run, the validity of Swan's independence result is sensitive to the production technology under consideration.

By pointing to the role of commitment ability of durable good producers, Bulow (1986) identifies another important reason why Swan's result does not

---

[2] There are other extensions of Swan's analysis. For example, Schmalensee (1974) and Su (1975) take into account maintenance effort of consumers, Ramm (1974) investigates the impact of imperfect capital markets on the independence result, and Rust (1986) and Hendel and Lizzeri (1999) explicitly consider second-hand markets. But these extensions are not relevant for our subsequent analysis and, hence, are not described in detail here.

hold, in general. Commitment ability means that the producer is able to credibly promise consumers that in future periods, she will not deviate from the plans announced today. Bulow shows that without commitment ability, a selling durable good monopolist has an incentive for planned obsolescence while a renting durable good monopolist has not. The argumentation is as follows. Bulow agrees with Swan's argument that profits are independent of whether the monopolist sells or rents the durable goods. Nevertheless, without commitment ability the selling durable good firm has an incentive to supply more durable goods tomorrow than she announces today since the capital loss in the future stock of the durable good (the reduction in the value of the future stock due to the additional supply) is born by consumers. Rational consumers anticipate the future behavior of the selling monopolist and adjust the price they are willing to pay for the durable good today. In order to compensate the accompanying reduction in revenues, the monopolist pursues a strategy of planned obsolescence, i.e. she enhances future sales of the durable good by reducing product durability. In contrast, a renting durable good monopolist is the owner of the products and, hence, she bears future losses in the value of the durable stock and has no incentive for planned obsolescence even if she does not possess commitment ability.

Bulow (1986) also inquires into durable goods oligopoly. He shows that, besides the incentive for planned obsolescence, oligopolistic sellers without commitment ability have an additional incentive to distort product durability. They compete not only for the consumers' current willingness-to-pay for the services of the durable good, but also for the consumers' future willingness-to-pay for the services of the durable good because a part of current production is still used by consumers in future periods. Hence, in order to early snatch away some future willingness-to-pay from competitors, the durable good seller makes her products more durable and so, ceteris paribus, increases her own revenues while decreasing the revenues of her competitors (revenue raising effect of product durability). This incentive for increasing durability works in the opposite direction than the incentive for planned obsolescence. Therefore, in a selling durable goods oligopoly without precommitment ability of producers, the net effect of the opposing incentives may be that products last longer than under perfect competition. In sum, Bulow (1986) shows that Swan's independence result is true in rental markets or in sales markets with commitment ability of producers. In sales markets without precommitment ability, however, imperfectly competitive producers have incentives for both planned obsolescence and revenue raising so that product durability depends on market structure, in general.

Extending Previous Durable Good Models

All studies discussed above completely ignore environmental aspects of product durability. In order to fill this gap, the present study will extend previous durable good models mainly in three directions:

(1) *Taking Into Account Production Emissions of Consumer Durables*

As in case of nondurable goods, the production process of durable goods generates emissions like smoke, wastewater and solid waste. To analyze the implications of these production emissions for environmental policy, we first present the model of Goering and Boyce (1999) in Sect. 3.1. These authors inquire into environmental policy in a durable good oligopoly with pollutants arising in the production process and depending on both output and product durability. Such an emissions function is motivated by the fact that an increase in output or in product durability typically requires more material inputs in production and, thus, creates additional production emissions.

Our reexamination of the model of Goering and Boyce (1999) will reveal some technical problems which restrict the insights from the model with respect to environmental policy for consumer durables. Section 3.2 therefore modifies the model in order to render the policy analysis richer. More specifically, we generalize the production cost function and consider an emissions function which depends on output only. Such an emissions function is realistic for durable goods whose lifetime may be prolonged without additional material inputs, for example, if product durability is enlarged by replacing one material against another more durable one or by changing the construction of the product while keeping the amount of material inputs constant.

It is interesting to note that even in such a setting, prolonging the product's lifetime (indirectly) influences the amount of production emissions: With respect to the stock of the durable good, output and product durability are substitutes. Keeping the durable stock constant and increasing product durability therefore implies that output declines with the consequence that production emissions are reduced. To illustrate this *emissions-reducing effect* of durability, consider the example of automobiles and suppose that the stock of automobiles is normalized to one. If we consider a time span of 10 years, then the stock can be generated either by two cars running 5 years each or by one car with a lifetime of 10 years. Since production emissions of a 10 years car are assumed to be the same as of a 5 years car (the emissions function is independent of durability), the option of producing one 10 years car halves production emissions compared to the option of producing two 5 years cars.[3]

(2) *Taking Into Account Solid Consumption Waste of Consumer Durables*

After their useful life, consumption goods turn into consumption waste. This is an important aspect especially for durable consumption goods since products like e.g. cars or refrigerators possess large weight and volume and therefore generate significant amounts of solid waste. To the best of my knowledge, the

---

[3] The emissions-reducing effect of durability is still present if durability is an argument of the emissions function, too, and if an increase in durability leads to an underproportional increase in the amount of production emissions.

economic literature did not yet raise the issue of efficient solid waste management for consumer durables. Chapter 4 of the present study will therefore develop a durable good model in which consumption goods turn into consumption waste after they are taken out of use. The consumption waste causes disposal and environmental costs.

In this setting, increasing product durability has two effects on solid consumption waste. First, for constant output of the durable good, enlarging product durability has a *waste-delaying effect* in the sense that some units of the output wear out at a later date. For example, suppose a producer manufactures 100 cars from which 60 cars wear out during the first five years and 40 cars wear out during the second five years of the output's lifetime. If (average) product durability of these 100 cars is increased, then the amount of scrapped units decreases during the first and second five years, say to 50 and 30 units, but increases during the third five years of the output's life from 0 to 20 units. Hence, making the goods more durable delays waste generation while leaving unaltered the overall amount of waste. Second, for constant stock of the durable good, enlarging product durability exerts a *waste-reducing effect* since output of the durable good declines. To illustrate this effect, consider again the example of 5 and 10 years cars. If the material needed for a 5 years car is more than half the weight of a 10 years car, then the amount of waste from a 10 years car falls short of the amount of waste from two 5 years cars.

(3) *Taking Into Account Product Recyclability*

As already outlined above, there is a recent line of environmental economic literature inquiring into the economics of recycling characteristics of consumption goods. This is done without explicit reference to product durability. Nevertheless, a closer look at the process of product design reveals that there is a link between product durability and product recyclability since both attributes are related to product weight: On the one hand, increasing weight may enhance recyclability since each unit of the durable contains more material and, thus, may facilitate the recyclers' effort to recover the material embodied in worn-out products. On the other hand, the increase in weight may also enhance product durability since owing to the additional material the good's decay process is retarded. The empirical significance of this relationship between durability and recyclability can easily be illustrated by referring to the example of aluminum car bodies. One reason for short lifetimes of aluminum car bodies are some mechanical properties of aluminum, in particular its relatively low rigidity. Carle and Blount (1999) argue that it is possible to compensate for lower rigidity by increasing wall thickness. Hence, increasing the quantity of material enlarges not only product weight, but also improves product durability. In addition, the generation of recycled material can be expected to increase since more mass of aluminum is available.

The relationship between product durability and recyclability will be investigated in Chap. 5 of our analysis. To fix our ideas, we first suppose product durability to be exogenously given, but then introduce a functional relation between durability and recyclability where, for simplicity, recyclability is assumed to be identical to product weight. The functional relation states that increasing recyclability renders the consumption goods more durable.

It should be noted that our analysis focuses on emissions arising during production or after consumption of durable goods. The analysis ignores emissions arising during the use of the products (as, for example, CO_2 emissions from driving a car) although such emissions are empirically significant in some cases. However, the reason for neglecting such emissions is that they are typically not affected by product durability, but by the stock of the durable good and by other product characteristics like e.g. fuel consumption.[4]

Questions to be Investigated in the Extended Models

Having described how the present study extends previous durable good models, we now clarify in more detail the economic questions to be investigated in the extended models. As already observed, our main focus is on environmental policy in durable good markets. In doing so, we will proceed stepwise:

(1) *Allocative Efficiency*

As a point of reference, we will first derive the efficient allocation. Chapters 3 and 4 employ partial equilibrium models since they deal not only with perfectly competitive markets, but also consider durable goods monopoly and oligopoly. We follow the usual procedure in partial equilibrium models to measure efficiency by social welfare which in our models equals consumer benefits from the (services of the) durable good less the sum of production, environmental and disposal costs. Chapter 5 uses a general equilibrium model with a representative consumer. In this setting, the efficient allocation is characterized by maximizing the representative consumer's utility taking into account the pertinent technical restrictions and resource constraints. Whatever the procedure, we mainly aim to answer two questions in deriving the efficient allocation: First, how is the efficient allocation characterized and, second, how is efficient durability influenced by environmental, disposal or recycling costs?

(2) *Market Equilibrium and Laissez Faire*

To make the case for environmental policy, one needs to establish market failure in a laissez faire economy. For this purpose, we consider a market economy with profit-maximizing firms and utility-maximizing consumers in the absence of taxes and markets for environmental assets (Chaps. 3 and 4) or markets for product design variables (Chap. 5). Chapters 3 and 4 consider the market

---

[4] There is a relation between durability and emissions arising during the use of cars if we account for technical progress in form of learning effects. The conclusion discusses this relation as an aspect for future research.

solution under both perfect and imperfect competition while Chap. 5 focuses on perfect competition only. By comparing the laissez faire allocation to the efficient allocation it is then possible to identify and specify market failure. It will be of special interest whether the various market imperfections (environmental externality, product design externality, imperfect competition) distort product durability and, if so, whether an underprovision or an overprovision of durability results.

(3) First-Best Regulation

Having identified and specified market failure under laissez faire, we proceed by deriving regulatory policies capable of restoring efficiency in the market economy. To this end, we introduce a particular set of tax rates and show how these tax rates have to be set in order to render the market equilibrium efficient. Of particular interest is whether Pigouvian regulation is sufficient for efficiency or whether the various market imperfections make it necessary to complement or even replace the Pigouvian approach.

(4) Second-Best Regulation

If the regulator needs at least two non-zero tax rates to restore efficiency of the market equilibrium and if she does not have at her disposal all of these tax rates, then the question of second-best taxation arises, namely how to set the available tax rates in order to maximize social welfare. Second-best emissions taxation under imperfect competition will be of particular interest in the subsequent analysis. According to the conventional economic intuition presented e.g. in Baumol and Oates (1988, pp. 79), the second-best emissions tax rate under imperfect competition falls short of the Pigouvian level, i.e. the tax rate is smaller than marginal environmental costs (underinternalization): Taxing emissions at a rate equal to marginal environmental costs completely internalizes the environmental externality, indeed, but output is inefficiently low owing to the firms' market power. An increase in the tax beyond marginal environmental costs would cause a welfare loss. It renders the firms' activities more expensive and, thus, tends to further reduce output. In contrast, lowering the tax below marginal environmental costs is welfare-enhancing since it induces firms to step up their production towards the efficient output level. This intuition for underinternalization has formally been proved under several restrictions in monopoly (Barnett, 1980) and oligopoly (Ebert, 1992). However, a second-best emissions tax exceeding marginal environmental costs (overinternalization) cannot be ruled out, in general. The literature demonstrates the possibility of such cases of overinternalization in nondurable goods oligopoly with a rather general emissions function (Ebert, 1992), with asymmetric firms (Simpson, 1995) or with free entry (Katsoulacos and Xepapadeas, 1995; Requate, 1997). Goering and Boyce (1997) even show that the second-best emissions tax in a durable goods monopoly may be greater than marginal environmental costs. However, they assume an exogenously given product durability and exclusively focus on the monopoly case. In contrast, the subsequent analysis will reexamine second-best emissions taxation in a durable goods oligopoly with endogenous product durability.

All issues raised above are related to environmental policy for consumer durables. Since our models build on previous durable good studies which inquire into the validity of Swan's independence result, we will also check whether this result remains true in our extended models. This is of special interest since the subsequent analysis takes into account several tax rates which are not considered in the previous durability literature and which may influence the validity of Swan's independence result.

Definitions, Assumptions and Basic Model

Before analyzing environmental policy for consumer durables, it proves helpful to introduce some basic concepts. In this chapter, we therefore present and discuss definitions, assumptions and mathematical tools which are required in the subsequent chapters. More specifically, Sect. 2.1 clarifies the difference between durable and nondurable consumption goods (Sect. 2.1.1) and shows how a durable good (Sect. 2.1.2) and different vintages of a durable good (Sect. 2.1.3) may be modeled. Section 2.2 uses these modeling concepts and introduces a basic oligopoly model without environmental pollution. The model is mainly presented for technical purposes. Section 2.2.1 describes the model, develops a durable good oligopoly game and derives the market equilibrium. Section 2.2.2 clarifies how to formalize the stationary state of the market equilibrium. Section 2.2.3 carries out a comparative static analysis of the long run market equilibrium and, in doing so, discusses and clarifies different procedures used in previous durable good studies.

2.1 Modeling Consumer Durables

2.1.1 Durable versus Nondurable Goods

Typical examples for nondurable consumption goods are food products or one-way packaging. Buildings, cars, electronic applications or refillable beverage containers are examples for durable consumption goods. In a non-formal way, we introduce

Definition D1. *A nondurable good is a good which completely decays after the first use. A durable good is a good which may be utilized several times.*

This definition can be illustrated by the example of beverage containers. An one-way can is only filled once and discarded after it is drained the first time. Consequently, an one-way can is a nondurable good. In contrast, a refillable bottle is a durable good since it can be filled several times.

In order to formalize durable goods, we introduce the concept of the *use potential* Γ of a consumption good. Γ is the maximum number of services which the consumption good is able to provide to the user during its lifetime. For example, in case of a beverage container, Γ equals the number of fillments and in case of a car, Γ measures the total number of miles the car runs during its life. The use potential is closely related to the *lifetime L* of the good. According to Bellmann (1990) and Conn (1977), the product's lifetime measures the time span from the first use to the last use of the good. Define the product's *use intensity* Υ as the amount of services the durable provides to the user during a particular time period, e.g. miles per year, and assume that Υ is constant over the whole lifetime of the good. We then obtain the relation

$$L = \frac{\Gamma}{\Upsilon} \, .$$

For example, suppose a car has an use potential of 100.000 miles and the owner of the car drives 10.000 miles per year. Then $\Gamma = 100.000$ miles, $\Upsilon = 10.000$ miles per year and the car's lifetime amounts to $L = 10$ years. Throughout the subsequent analysis it is assumed that the use intensity Υ is fixed and, thus, the lifetime of the consumption good is proportional to the use potential of the good. Furthermore, we normalize Υ to unity with the consequence that a unit of the durable good provides exactly one service unit in every period of its life. The lifetime of the consumption good is then equal to the use potential.

All other things being equal, the product's use potential or lifetime, for that matter, is determined by several factors under the control of producers. Examples are given by Jambor and Beyer (1997, p. 205):

- *Choice of material inputs*: Materials differ with respect to their durability, rigidity or decay. The choice of materials in the production process therefore influences the use potential of the product, i.e. how often or how long the product can be used. For example, beverage containers manufactured from glass are subject to higher risk of breaking or scratching than plastic bottles, especially PET bottles. Car bodies are another example. In recent years, an increasing number of cars are manufactured with full aluminum bodies since aluminum is less susceptible to corrosion than steel.

- *Construction of the durable good*: Durability, rigidity and decay are not only influenced by the material employed in the production process, but also by the construction or the design of the product. For example, a car body with relatively large wall-thickness possesses relatively high rigidity and usually lasts longer and has greater use potential than a car body with relatively thin walls.

- *Processing of the product*: Use potential and lifetime of a product are also influenced by the processing of the material input in the production process. For example, if a car is not processed with due care, then it is not robust and breaks down already after a few thousand miles.

The subsequent analysis proceeds on the assumption that all these factors can be summarized in a single product design variable $\phi \in \mathbb{R}_+$ which is called *built-in product durability*.

Figure 2.1 sums up the basic relations to be used in our analysis. The

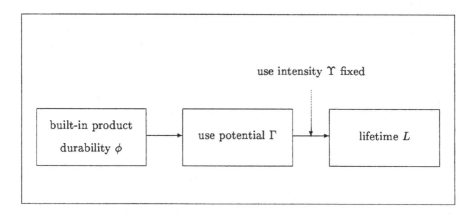

Fig. 2.1. Product Durability, Use Potential and Lifetime of a Consumption Good

producer chooses built-in product durability ϕ which influences the product's use potential Γ. Since the use intensity Υ of the product is constant and normalized to unity, the product's lifetime equals the product's use potential and is therefore also influenced by built-in product durability.

2.1.2 Decay Function of Consumer Durables

Until now we have focused on individual product units. In general, a producer manufactures many units of a durable good. Let y be the output of the durable good. If all units of y are produced in the same period, on the same production plant and with the same technology, then they possess (almost) the same built-in product durability ϕ. However, we cannot expect that also the lifetime is the same for all units. Besides built-in product durability, the product's lifetime usually depends on additional factors not under the control of producers. Examples for such factors are maintenance effort of consumers, random material failure during the use of the good or risk of accident. In formal terms, we write the product's lifetime as a function $L(\phi, \mathbf{z})$ where the vector $\mathbf{z}$ comprises all determinants of the lifetime not under the control of producers. Although in reality the vector $\mathbf{z}$ may be influenced by consumers and contains both deterministic and random variables, the subsequent analysis simplifies and proceeds on the assumption that $\mathbf{z}$ is exogenously given and deterministic.[1] Nevertheless, different units of the durable good then still

---

[1] In case that $\mathbf{z}$ is a random vector, we have to consider stochastic models in which households and firms act under uncertainty and maximize expected utility and

possess different values of $\mathbf{z}$ and, thus, different lifetimes even if they have the same built-in product durability. Therefore, some units of y are scrapped at relatively low age whereas others wear out at relatively high age.

Such a decay process is conveniently modeled by the so-called decay function employed, for example, by Sieper and Swan (1973), Abel (1983) and Goering (1993). In a nutshell, the decay function describes how the units of the durable output wear out over time and how this process is influenced by the built-in product durability. More formally, the decay function is a function

$$D: \mathbb{R}_+^2 \mapsto [0, 1], \tag{2.1}$$

where $D(a, \phi)$ can be interpreted in two alternative ways. The first interpretation is to consider the physical units of the durable good and to denote by $D(a, \phi)$ the share of y still available for use when the units are a years old. Correspondingly, $1 - D(a, \phi)$ is the share of y already scrapped at age a and $-D_a(a, \phi)$ represents the scrapping rate at age a, i.e. the share of y which wears out at age a. It follows that, when output y has been produced and sold a years ago, then $yD(a, \phi)$ units are still in use, $y[1 - D(a, \phi)]$ units have already been discarded and $-yD_a(a, \phi)$ units wear out in the present year and are no longer available thereafter. The second interpretation of the decay function (2.1) suggests to consider the services provided by the output. Since the use intensity of the durable good is normalized to unity (with the consequence that every unit of the durable good provides exactly one service unit per period), $yD(a, \phi)$ represents the remaining use potential of y at age a, $y[1 - D(a, \phi)]$ measures the use potential of y which is already used up at age a and $-yD_a(a, \phi)$ gives the loss in the use potential of y at age a. Throughout the subsequent analysis we use both interpretations of the decay function interchangeably.[2]

The next step is to specify some (plausible) properties of the decay function. We will first impose a set of rather general assumptions and then consider some specific functional forms of D. It will also be clarified to which extent these functional forms satisfy the assumptions imposed on the decay function.

Properties of the Decay Function

One basic property of the decay function already specified in (2.1) is that its range is the interval $[0, 1]$. Hence, the available share of output y has to be

---

expected profits, respectively. This would complicated the analysis without adding further insights into environmental policy for consumer durables.

[2] A third interpretation is possible if we think of decay as a stochastic process. $-D_a(a, \phi)$ is then the density function of product lifetime. $D(a, \phi)$ is the associated distribution function measuring the probability that the lifetime of a product with durability ϕ is not greater than a. Analogously, $1 - D(a, \phi)$ is the probability that a product with durability ϕ has a lifetime greater than a. As already explained above, however, we avoid taking into account uncertainty and assume a deterministic decay process.

non-negative and not greater than one due to physical constraints. Moreover, for D to be a meaningful decay function, we introduce

Assumption A1. *The decay function (2.1) is twice continuously differentiable and satisfies*

$$D(0,\phi) = 1 \quad and \quad \lim_{a \to \infty} D(a,\phi) = 0 \qquad \forall\, \phi \in \mathbb{R}_+ , \tag{2.2}$$

$$D_a(a,\phi) \leq 0 \qquad \forall\, (a,\phi) \in \mathbb{R}_+^2 , \tag{2.3}$$

$$D_\phi(a,\phi) \geq 0 \qquad \forall\, (a,\phi) \in \mathbb{R}_+^2 , \tag{2.4}$$

$$\exists\, (\tilde{a}, \tilde{\phi}) \in \mathbb{R}_+^2 \quad such\ that \quad D_\phi(\tilde{a}, \tilde{\phi}) > 0 . \tag{2.5}$$

The decay process of the durable good is expressed by (2.2) and (2.3). The first part of (2.2) states that at the production time (at age zero), all units of the output are still in use. The second part of (2.2) says that the amount of available units tends to zero as age grows without bounds. Note that this assumption is also satisfied if all units of the durable good wear out in finite time. According to (2.3), the available share of output is non-increasing in age a and, in connection with (2.2), strictly decreases at least during some age intervals. The remaining conditions in assumption A1 describe the impact of built-in product durability on the decay process. According to (2.4) and (2.5), shifting built-in durability to a higher level never decreases the available output share, but increases the share at least for some ages. Roughly speaking, assumption A1 states that output decays over time until all units are scrapped and that increasing built-in product durability retards this decay process. A1 represents a minimum set of assumptions which plausible decay functions ought to satisfy and we therefore presume assumption A1 throughout all subsequent chapters.

In some parts, an additional assumption on the decay function is applied. This assumption specifies *how* changes in built-in product durability influence the decay rate of the durable good at different ages of the output.

Assumption A2. *The decay function (2.1) satisfies*

$$\forall\, \phi \in \mathbb{R}_+ \quad \exists\, A(\phi) \quad with \quad D_{a\phi}(a,\phi) \gtreqless 0 \quad \Leftrightarrow \quad a \lesseqgtr A(\phi) . \tag{2.6}$$

To interpret this assumption, note that $-D_{a\phi}(a,\phi)$ represents the change in the scrapping rate at age a due to a change in product durability. An increase in product durability prolongs the lifetime of some units of output so that these units wear out at a later date. According to (2.6), this prolongation decreases the scrapping rate at relatively low age $a < A(\phi)$ and increases it at relatively high age $a > A(\phi)$. In terms of service units, increasing product durability reduces the loss in the use potential of output in early periods $a < A(\phi)$ and raises it in later periods $a > A(\phi)$.

Assumptions A1 and A2 have some interesting implications which turn out to be useful in the subsequent chapters. To derive these implications, let

$$\Gamma(\phi) := \int_0^\infty D(a, \phi)da , \qquad \Psi(\phi) := \int_0^\infty D(a, \phi)e^{-\delta a}da ,$$

$$\Lambda(\phi) := -\int_0^\infty D_a(a, \phi)da , \qquad \Omega(\phi) := -\int_0^\infty D_a(a, \phi)e^{-\delta a}da ,$$

with $\delta \in \mathbb{R}_{++}$. Even though these definitions appear quite technical, they have economically meaningful interpretations. $y\Gamma(\phi)$ is the total amount of services (the total use potential) which output y provides over its whole lifetime. Similarly, if we interpret δ as discount rate, then $y\Psi(\phi)$ is the total discounted use potential of output y. $y\Lambda(\phi)$ gives the sum of all changes in the use potential or, in physical units, the total amount of scrapped units during the life of output y. $y\Omega(\phi)$ is the discounted counterpart. With the help of these definitions, we obtain

Lemma 2.1. (i) *If the decay function (2.1) satisfies A1, then*

$$\Gamma(\phi) > 0 , \quad \Psi(\phi) > 0 , \quad \Lambda(\phi) = 1 , \quad \Omega(\phi) > 0 ,$$

$$\Gamma'(\phi) > 0 , \quad \Psi'(\phi) > 0 , \quad \Lambda'(\phi) = 0 ,$$

$$\Gamma'(\phi)/\Gamma(\phi) > \Psi'(\phi)/\Psi(\phi) ,$$

for all $\phi \in \mathbb{R}_+$.
(ii) *If the decay function (2.1) satisfies A1 and A2, then it is also true that*

$$\Omega'(\phi) < 0 ,$$

for all $\phi \in \mathbb{R}_+$.

Proof. $\Gamma(\phi), \Psi(\phi), \Omega(\phi) > 0$ follows from the differentiability of D and from $D(a, \phi) \in [0, 1]$, (2.2), (2.3) and $e^{-\delta a} \in [0, 1]$. Using (2.2) yields

$$\Lambda(\phi) = -\int_0^\infty D_a(a, \phi)da = -\Big[D(a, \phi)\Big]_{a=0}^{a=\infty} = D(0, \phi) - \lim_{a \to \infty} D(a, \phi) = 1 .$$

$\Lambda'(\phi) = 0$ follows directly from $\Lambda(\phi) = 1$. (2.4), (2.5) and $e^{-\delta a} \in [0, 1]$ imply

$$\Gamma'(\phi) = \int_0^\infty D_\phi(a, \phi)da > 0 , \qquad \Psi'(\phi) = \int_0^\infty D_\phi(a, \phi)e^{-\delta a}da > 0 .$$

Moreover, owing to the generalized mean value theorem there exists a function $\widetilde{A} : \mathbb{R}_+ \mapsto \mathbb{R}_+$ with the image $\widetilde{A}(\phi)$ such that

$$\Psi(\phi) = \int_0^\infty D(a, \phi)e^{-\delta a}da = e^{-\delta\widetilde{A}(\phi)} \int_0^\infty D(a, \phi)da = e^{-\delta\widetilde{A}(\phi)}\Gamma(\phi) .$$

$\tilde{A}$ is a function of ϕ since the integrand D depends on ϕ. Owing to (2.4) and (2.5) we have $D_\phi(a, \phi) > 0$ at least for one $(a, \phi) \in \mathbb{R}_+$ and thus $d\tilde{A}(\phi)/d\phi > 0$. The above expression for Ψ then implies

$$\frac{\Psi'(\phi)}{\Psi(\phi)} = \frac{\Gamma'(\phi)}{\Gamma(\phi)} - \delta \frac{d\tilde{A}(\phi)}{d\phi} < \frac{\Gamma'(\phi)}{\Gamma(\phi)} .$$

This completes the proof of Lemma 2.1i. To show Lemma 2.1ii, note that the derivative of Ω may be written as

$$\Omega'(\phi) = -\int_0^\infty D_{a\phi}(a, \phi)e^{-\delta a}da$$

$$= -\int_0^{A(\phi)} D_{a\phi}(a, \phi)e^{-\delta a}da - \int_{A(\phi)}^\infty D_{a\phi}(a, \phi)e^{-\delta a}da .$$

We apply the generalized mean value theorem to both integrals of this expression. Then there exist two functions $\bar{A} : \mathbb{R}_+ \mapsto \mathbb{R}_+$ and $\hat{A} : \mathbb{R}_+ \mapsto \mathbb{R}_+$ with the images $\bar{A}(\phi) \in]0, A(\phi)[$ and $\hat{A}(\phi) \in]A(\phi), \infty[$ such that

$$\Omega'(\phi) = -e^{-\delta\bar{A}(\phi)} \int_0^{A(\phi)} D_{a\phi}(a, \phi)da - e^{-\delta\hat{A}(\phi)} \int_{A(\phi)}^\infty D_{a\phi}(a, \phi)da .$$

Due to assumption A2 and $\Lambda'(\phi) = 0$ we may write $B(\phi) := \int_0^{A(\phi)} D_{a\phi}(a, \phi)da$ $= -\int_{A(\phi)}^\infty D_{a\phi}(a, \phi)da > 0$. Inserting $B(\phi)$ in $\Omega'(\phi)$ yields

$$\Omega'(\phi) = B(\phi) \left[e^{-\delta\hat{A}(\phi)} - e^{-\delta\bar{A}(\phi)} \right] .$$

Since $\bar{A}(\phi) < \hat{A}(\phi)$, we obtain $e^{-\delta\hat{A}(\phi)} < e^{-\delta\bar{A}(\phi)}$ and $\Omega'(\phi) < 0$. $\qquad\square$

Lemma 2.1 may be interpreted as follows. $\Gamma(\phi), \Gamma'(\phi), \Psi(\phi), \Psi'(\phi) > 0$ state that both the undiscounted use potential Γ and the discounted use potential Ψ of output are positive and increasing in built-in product durability ϕ. This is intuitively plausible since increasing durability prolongs the lifetime of some units of the durable good without reducing the lifetime of any other unit. The units tend to be longer in use, provide more services and increase the use potential of output. According to $\Gamma'/\Gamma > \Psi'/\Psi$, the relative change in the undiscounted use potential exceeds the relative change in the discounted use potential. For interpreting Λ, Ω and their derivatives, consider the physical units of output. Since we have assumed that the decay function tends to zero as age grows without bounds, all units of output sooner or later wear out. Consequently, the total amount of scrapped units, $y\Lambda(\phi)$, has to equal the quantity of the durable good produced, y, so that we obtain $\Lambda(\phi) = 1$. Of course, this is true for any value of the built-in product durability: Increasing durability prolongs the lifetime of some units of output, indeed, but these units are not in use forever. They wear out in some future period and, hence, the total amount of scrapped units is left unaltered, i.e. $\Lambda'(\phi) = 0$. If the impact of durability on the scrapping rates is described by A2, then raising durability reduces the amount of scrapped units $-y D_a(a, \phi)$ at relatively low age, but increases it at relatively high age. If we weigh the changes in the scrapping rates with the declining factor $e^{-\delta a}$, then the decreasing amounts of scrapped units at relatively low age get larger weight than the increasing amounts of scrapped units at relatively high age and the sum of the (weighted) changes in the scrapping rates becomes negative, i.e. $\Omega'(\phi) < 0$.

Exponential Decay Function

The first parametric decay function to be discussed is the exponential function which has been suggested e.g. by Muller and Peles (1990). It is defined by

$$D(a, \phi) = e^{-a/\phi} . \tag{2.7}$$

(2.7) satisfies assumption A1 since it is twice continuously differentiable and

$$D(a, \phi) = 1 , \qquad \lim_{a \to \infty} D(a, \phi) = 0 ,$$

$$D_a(a, \phi) = -\frac{1}{\phi} e^{-a/\phi} < 0 , \qquad D_\phi(a, \phi) = \frac{a}{\phi^2} e^{-a/\phi} > 0 .$$

It also satisfies assumption A2 since there is $A(\phi) = \phi$ such that

$$D_{a\phi}(a, \phi) = \frac{e^{-a/\phi}}{\phi^2} \left[1 - \frac{a}{\phi} \right] \gtreqqless 0 \quad \Leftrightarrow \quad a \lesseqqgtr A(\phi) .$$

Hence, Lemma 2.1 holds for the exponential decay function (2.7). More specifically, we obtain

$$\Gamma(\phi) = \phi > 0 , \quad \Psi(\phi) = \frac{\phi}{1 + \delta\phi} > 0 , \quad \Lambda(\phi) = 1 , \quad \Omega(\phi) = \frac{1}{1 + \delta\phi} > 0 ,$$

$$\Gamma'(\phi) = 1 > 0 , \quad \Psi'(\phi) = \frac{1}{(1 + \delta\phi)^2} > 0 ,$$

$$\Lambda'(\phi) = 0 , \quad \Omega'(\phi) = -\frac{\delta}{(1 + \delta\phi)^2} < 0 .$$

As an illustration of the properties of the exponential decay function, consider the plots in Fig. 2.2. The left panel shows the graph of the decay function for alternative degrees of product durability. This plot illustrates that the exponential decay function satisfies assumption A1: The available share of output is one at $a = 0$, it decreases with increasing age and decreasing product durability and it tends to zero as the age becomes arbitrarily large. The right panel

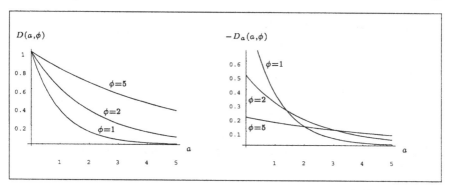

Fig. 2.2. Exponential Decay Function $D(a, \phi) = e^{-(a/\phi)}$

of Fig. 2.2 shows the scrapping rate as a function of age for alternative product durabilities. This plot illustrates two properties of the exponential decay function. First, it can be seen that the exponential decay function satisfies assumption A2: An increase in product durability decreases the scrapping rate at relatively low age and increases the scrapping rate at relatively high age. Second, the scrapping rate is strictly decreasing in age for any given product durability. This property is not presupposed in assumptions A1 and A2. Indeed, it is not needed for our subsequent analysis. But it is worth to note in passing that for many durable goods, this property of the exponential decay function appears somewhat unrealistic: If the scrapping rate strictly decreases with age, then a relatively large part of output is already scrapped in the early time of the output's life while in later periods only relatively few units wear out. At best, this might be descriptive for newly introduced products that are very unreliable and susceptible to failure at relatively low age already. However, for most durable goods, we expect that at relatively low and high ages only few units fail while the major part is scrapped at intermediate age.[3] It is therefore more realistic that the scrapping rate increases at low ages and decreases after a certain point in time. We will now consider another parametric functional form of the decay function which possesses this additional property.

Generalized Exponential Decay Function

In a more general way, the exponential decay function may take the form

$$D(a, \phi) = e^{-(a/\phi)^{\vartheta}} \qquad \text{with} \qquad \vartheta \geq 1 \,. \qquad (2.8)$$

Obviously, the decay function (2.7) is a special case of (2.8) for $\vartheta = 1$. The function (2.8) has the properties $D(0, \phi) = 1$, $\lim_{a \to \infty} D(a, \phi) = 0$ and

---

[3] An empirical example is provided by Table 4 in OECD (1982) which depicts the scrapping rates of different vintages of cars.

$$D_a(a, \phi) = -\vartheta \frac{a^{\vartheta-1}}{\phi^\vartheta} e^{-(a/\phi)^\vartheta} < 0 , \qquad D_\phi(a, \phi) = \vartheta \frac{a^\vartheta}{\phi^{\vartheta+1}} e^{-(a/\phi)^\vartheta} > 0 ,$$

$$D_{a\phi}(a, \phi) = \vartheta^2 \frac{a^{\vartheta-1}}{\phi^{\vartheta+1}} \left[1 - \left(\frac{a}{\phi} \right)^\vartheta \right] e^{-(a/\phi)^\vartheta} \gtreqless 0 \quad \Leftrightarrow \quad a \lesseqgtr \phi .$$

Hence, it satisfies the assumptions A1 and A2 and Lemma 2.1 can be applied to this function, too. The distinctive feature of the generalized exponential decay function is that the scrapping rate $-D_a(a, \phi)$ is not necessarily decreasing in age since

$$D_{aa}(a, \phi) = \vartheta \frac{a^{\vartheta-2}}{\phi^\vartheta} \left[1 - \vartheta + \vartheta \left(\frac{a}{\phi} \right)^\vartheta \right] e^{-(a/\phi)^\vartheta} \lesseqgtr 0$$

$$\Leftrightarrow \quad a \lesseqgtr \phi[(\vartheta - 1)/\vartheta]^{1/\vartheta} =: \tilde{a} .$$

Clearly, for $\vartheta = 1$, the scrapping rate $-D_a(a, \phi)$ is strictly decreasing in age, but for $\vartheta > 1$, the scrapping rate increases for relatively low ages $a < \tilde{a}$ and decreases for relatively high ages $a > \tilde{a}$. This property of the generalized exponential decay function is illustrated in Fig. 2.3 for $\vartheta = 2$. The left panel

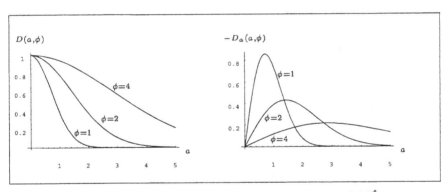

Fig. 2.3. Generalized Exponential Decay Function $D(a, \phi) = e^{-(a/\phi)^\vartheta}$ with $\vartheta = 2$

captures the generalized exponential decay function for alternative product durabilities. In contrast to the exponential decay function (2.7), the generalized function is not convex at all ages, but has a concave segment for relatively low ages. This property is also reflected in the right plane of Fig. 2.3 which shows the scrapping rate as a function of age for alternative product durabilities. Evidently, the curve of the scrapping rate is bell shaped. It increases up to a certain age and declines thereafter. Hence, at relatively low and high ages, only few units of the output wear out whereas the major part of the output becomes useless at intermediate age. Increasing product durability flattens the bell curve. It decreases the scrapping rate at relatively low ages and increases it at relatively high ages. As already argued above, these properties of the generalized exponential decay function appear realistic for many durable goods. It should be noted, however, that the subsequent analysis does not rely on the bell shape of the scrapping rate; the more specific exponential function (2.7) represents an appropriate decay function, too.

One-Hoss Shay Decay Function

Some of the early durable good studies as, for example, Levhari and Srinivasan (1969) or Swan (1970) used the so-called one-hoss shay decay function. This function is defined by

$$D(a, \phi) = \begin{cases} 1 & \text{for all } \quad a \leq \phi \,, \\ 0 & \text{for all } \quad a > \phi \,. \end{cases} \tag{2.9}$$

The underlying idea is that all units of output wear out simultaneously after exactly ϕ years. For this decay function, product durability ϕ equals the lifetime of the durable good.

(2.9) does not satisfy assumptions A1 and A2 because it is not continuously differentiable. Since we will restrict our attention to models satisfying at least assumption A1, the subsequent analysis will not apply to durable goods of the one-hoss shay type. However, (2.9) can be approximated by the generalized exponential decay function (2.8) if the parameter ϑ is chosen sufficiently large. To see this, observe that for $\vartheta \to \infty$, the generalized exponential decay function (2.8) is turned into (2.9) since

$$\lim_{\vartheta \to \infty} D(a, \phi) = \lim_{\vartheta \to \infty} e^{-(a/\phi)^{\vartheta}} = \begin{cases} 1 & \text{for } a \in [0, \phi[\,, \\ 1/e & \text{for } a = \phi \,, \\ 0 & \text{for } a \in]\phi, \infty[\,. \end{cases}$$

The slope of the generalized exponential decay function at $a = \phi$ reads $\lim_{\vartheta \to \infty} D_a(a, \phi) = -\lim_{\vartheta \to \infty} \vartheta/e = -\infty$. These properties are illustrated in Fig. 2.4. The left plane shows the one-hoss shay decay function (2.9) for a durable good with lifetime $\phi = 3$. The right plane depicts the generalized exponential decay function (2.8) for $\phi = 3$ and alternative values of ϑ. For increasing ϑ, the shape of the generalized exponential decay function becomes similar to the one of the one-hoss shay decay function. Hence, the latter function is 'approximately' reflected by our subsequent analysis if we let $\vartheta \to \infty$ in (2.8).

Two-Period Decay Function

The decay functions discussed above are defined in continuous time. Some previous studies like e.g. Bulow (1986) or Goering (1992) employed a discrete-time framework for modeling durable goods. Although these authors do not

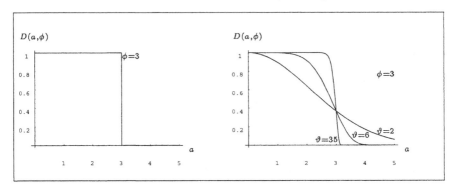

Fig. 2.4. One-Hoss Shay and Generalized Exponential Decay Function

formally introduce a decay function, it is easy to see that the decay process in their models follows the function $D : \mathbb{N} \times [0,1] \mapsto [0,1]$ with

$$D(a, \phi) = \begin{cases} 1 & \text{for} \quad a = 1 \,, \\ \phi & \text{for} \quad a = 2 \,, \\ 0 & \text{for} \quad a = 3, 4 \dots \,. \end{cases} \qquad (2.10)$$

Consider a producer who manufactures y units of a durable good. All units last at most two periods. Production takes place at the beginning of the first period $a = 1$ of the output's life. During this first period, all units are in use so that the decay function is equal to one. At the end of the first period, $(1-\phi)y$ units wear out while the remaining ϕy units carry over to the second period $a = 2$ of the output's life. Hence, in the second period, the decay function takes on the value ϕ. At the end of the second period of the output's life, also the remaining ϕy units are scrapped such that the decay function becomes zero if the age of the output is $a > 2$. In this setting, product durability ϕ measures that part of output which survives the first period and is still in use in the second period of the output's life.

Like the one-hoss shay decay function, the two-period decay function does not satisfy the assumptions A1 and A2. Moreover, the discrete-time approach prevents us from approximating (2.10) by another decay function satisfying A1 and A2. Nevertheless, the two-period decay function possesses similar properties as listed in assumption A1. It is non-increasing in age, non-decreasing in product durability, equal to unity in the first-period of the output's life and it becomes zero as the age of the output tends to infinity. Moreover, owing to its simple form, the two-period decay function (2.10) has the advantage of simplifying the modeling of durable goods so that it becomes possible to add complexity to the model elsewhere without rendering the analysis intractable. We will therefore make use of the two-period decay function in some parts of the subsequent analysis.

2.1.3 Vintage Durable Goods

Until now we have considered y units of a durable good with product durability ϕ produced at a particular point in time. During the future life of these units the producer may manufacture further units and perhaps she has already produced some units before the point in time under consideration. In general, units produced at different dates possess different degrees of product durability. To formally express this observation, we need to consider a vintage approach for the durable good. Let v be an index indicating vintages. The producer manufactures $y(v)$ units of vintage v with built-in product durability $\phi(v)$. The decay process of vintage v is described by the decay function $D[a(v), \phi(v)]$ where $a(v)$ is the age of vintage v. It should be noted that the decay function itself is not vintage-specific. Differences in the decay process of vintages are solely due to differences in output and built-in product durability.

Now consider a specific time t. The available share of vintage $v \leq t$ at this time is $D[t - v, \phi(v)]$ since the age of vintage v at time t reads $a(v) = t - v$. Hence, $D[t - v, \phi(v)]y(v)$ units of vintage v are still in use at time t. If we use a continuous time framework and add up the units of all vintages still in use at t, then we obtain

$$c(t) = \int_{-\infty}^{t} D[t - v, \phi(v)]y(v)dv . \tag{2.11}$$

$c(t)$ is the stock of the durable good consisting of all units available for use at time t. Since the use intensity of a unit of the durable good is normalized to unity, the stock of the durable good also measures the total use potential of the durable at time t (the total amount of services provided by the units which are still in use at time t).

Note that output $y(v)$ in (2.11) is a flow of goods whereas $c(t)$ represents a stock. This observation reveals the dynamic feature of the vintage durable good approach. Changing output and durability of a vintage v changes not only the stock of the durable good at time v, but also influences the stock of the durable good at times $t > v$. Moreover, (2.11) reveals that output and durability are substitutes with respect to the durable stock. For any given stock of the durable good at time t, increasing product durability of a particular vintage $\tilde{v} < t$ allows reducing output of this vintage or output of another vintage still in use at time t. This impact of durability will turn out to be important in our analysis of environmental policy in durable goods markets.

2.2 Basic Oligopoly Model without Pollution

2.2.1 Model Description and Market Equilibrium

With the help of the modeling concepts presented in the previous section, we will now develop a durable good model without pollution. The model is akin to the ones investigated in Sieper and Swan (1973) and Abel (1983). The difference to these studies is that our model is not restricted to the polar market structures of perfect competition and monopoly, but also allows for studying the case of durable goods oligopoly. Furthermore, while previous authors focus on the analysis of Swan's independence result, we use the basic model in order to present and discuss some basic mathematical tools typically employed in durable good studies. We will take advantage of these discussions in the subsequent analysis of durable goods economies with environmental pollution.

Model Description

Consider a durable good industry consisting of $n \geq 1$ identical firms. Let $N := \{1, \ldots, n\}$ be the set of all firms. The model is given by

$$k_i(t) = K[y_i(t), \phi_i(t)] \qquad \forall\, i \in N \,, \tag{2.12}$$

$$c_i(t) = \int_{-\infty}^{t} D[t - v, \phi_i(v)] y_i(v) dv \qquad \forall\, i \in N \,, \tag{2.13}$$

$$p(t) = P[C(t)] \qquad \text{with} \qquad C(t) = \sum_{j \in N} c_j(t) \,. \tag{2.14}$$

Equation (2.12) describes firm i's production costs at time t when it produces $y_i(t)$ units of the durable good with built-in product durability $\phi_i(t)$. $K : \mathbb{R}_+^2 \mapsto \mathbb{R}_+$ is the firm-independent cost function. With respect to this function, we introduce

Assumption A3. *The cost function $K(y, \phi)$ is twice continuously differentiable. It satisfies*

$$K_y(y, \phi) > 0 \,, \qquad K_\phi(y, \phi) > 0 \,, \tag{2.15}$$

$$K_{yy}(y, \phi) \geq 0 \,, \qquad K_{\phi\phi}(y, \phi) \geq 0 \,, \tag{2.16}$$

$$\frac{\partial}{\partial y} \left(\frac{\eta_{Ky}}{\eta_{K\phi}} \right) = 0 \,, \tag{2.17}$$

for all $(y, \phi) \in \mathbb{R}_+^2$ where $\eta_{Ky} := y K_y / K$ and $\eta_{K\phi} := \phi K_\phi / K$.

According to (2.15), the function K exhibits positive marginal costs of output and product durability. Firm i's production costs increase with increasing product durability because, for example, higher product durability requires more material input, more expensive materials or higher investment in durability research. Equation (2.16) states that the cost function is non-concave in output and product durability. With respect to product durability, this condition says that it does not become the easier to increase product durability the larger durability already is. Equation (2.17) represents the Abel condition already referred to in the introduction. Recall that Abel (1983) shows this condition to be sufficient for Swan's independence result. Even though the Abel condition seems to be restrictive at first glance, it is fulfilled by many cost functions often used in economic models. Examples are cost functions which are linear in output or, more generally, Cobb-Douglas cost functions. Further examples can be found in Abel (1983, p. 629).

$c_i(t)$ in (2.13) is firm i's stock of the durable good at time t. It comprises all units of the vintages $v \leq t$ still available for use at time t. Equation (2.14) represents the demand side of the model. $C(t)$ is the total industry stock of the durable good. It equals the total amount of services available at time t since the use intensity is hold constant and normalized to unity. We assume that consumers derive utility from the services of the durable good. Hence, the demand function $P : \mathbb{R}_+ \mapsto \mathbb{R}_+$ in (2.14) gives the price of a unit of the durable's services as a function of the total quantity of services $C(t)$. This price equals the rental price of the durable good and, therefore, does not directly depend on product durability. Such a formalization of the demand side supposes that the durable good is homogenous, i.e. every product unit provides consumers the same services independent of its producer and independent of its age. The demand function is assumed to satisfy

Assumption A4. *The demand function $P(C)$ is twice continuously differentiable. It satisfies*

$$P'(C) < 0 , \tag{2.18}$$

$$P'(C) + c_i P''(C) < 0 , \tag{2.19}$$

for all $C \in \mathbb{R}_+$ and for all $i \in N$.

According to (2.18), the demand function is downward-sloping so that an increase in the industry stock of the durable good reduces the price of the durable's services. Condition (2.19) is the usual stability condition imposed in oligopoly models, for example, by Dixit (1986). It states that firm i's marginal revenues $P(C) + c_i P'(C)$ declines with increasing industry stock of the durable.

Durable Goods Oligopoly Game

Since the demand function depends on the total industry stock of the durable good, revenues of the individual firm depend not only on the choice variables of this firm, but also on the choice variables of all other firms. Such an oligopolistic interdependence can be mapped by a durable goods oligopoly game. In this subsection, we formally describe the game (players, strategies, information pattern, pay-off functions) and specify the equilibrium concept of the game.

Let us start by describing the game. *Players* of the game are the n durable good firms. N is the set of all players. Player i chooses the time paths of output and product durability over the planning period $[0, \infty[$. For notational convenience, let $x(a, b) := \{x(t) \mid a \leq t \leq b,\, a, b \in \mathbb{R}\}$ be the path of variable x over the time interval $[a, b]$. The *strategy of player* i is then denoted by $s_i := [y_i(0, \infty), \phi_i(0, \infty)]$ where $y_i(0, \infty)$ and $\phi_i(0, \infty)$ are the time paths of firm i's output and product durability, respectively. Player i chooses his strategy from the *strategy space* $\mathcal{S} := \{[y(0, \infty), \phi(0, \infty)] \mid y(t), \phi(t) \geq 0 \; \forall t \in [0, \infty[\}$ which is assumed to be firm-independent and represents the set of all time paths with non-negative output and product durability. $s_{-i} := [s_1, \ldots, s_{i-1}, s_{i+1}, \ldots, s_n]$ denotes a vector containing the strategies of player i's competitors.

As argued in Sect. 2.1.3, the accumulation of the durable good is captured by a dynamic equation. As a consequence, the durable goods oligopoly game turns out to be also dynamic. For dynamic games, it has to be specified which information pattern player i faces. According to Fudenberg and Tirole (1991, pp. 130) or Fershtmann et al. (1992), we distinguish open-loop and closed-loop information pattern. The open-loop approach assumes that player i conditions its strategy choice only on the information it has at time zero, i.e. information about the time paths $y_j(-\infty, 0)$ and $\phi_j(-\infty, 0)$ for all $j \in N$. It irrevocably commits to the whole time path of its decision variables right at the beginning of the planning period. The closed-loop approach is based on the assumption that, at every time $t > 0$, player i takes into account the information it received about the whole time path of the variables up to t, i.e. information about the time paths $y_j(-\infty, t)$ and $\phi_j(-\infty, t)$ for all $j \in N$. The firm does not precommit, but recalculates the optimality of its decision in each point in time based on the additional information it received. For the time being, we suppose all players face open-loop information pattern so that s_i is the open-loop strategy of player i. The implications of the closed-loop information pattern will be discussed in the last paragraph of this section on pp. 31.

In order to complete the description of the oligopoly game, we formalize the pay-off functions of the players. Durable good firms may pursue alternative marketing policies. They either rent or sell their output. For the time being, we explicitly focus on the rental case and analyze the sales case also in the last paragraph of this section on pp. 31. If firm i rents the products, then its rental profits at time t equal rental revenues $p(t)c_i(t)$ less production costs $k_i(t)$. Discounting all rental profits with the common discount rate $\delta \in \mathbb{R}_{++}$ and using the definitions in (2.12)–(2.14), the *pay-off of player* i is a function $\Pi^{ri} : \mathcal{S}^n \mapsto \mathbb{R}$ where

$$\Pi^{ri}(s_i, s_{-i}) = \int_0^\infty \left[P\left[\sum_{j \in N} \int_{-\infty}^t D[t - v, \phi_j(v)] y_j(v) dv \right] \times \right.$$

$$\left. \times \int_{-\infty}^t D[t - v, \phi_i(v)] y_i(v) dv - K[y_i(t), \phi_i(t)] \right] e^{-\delta t} dt \quad (2.20)$$

is the present value of firm i's rental profits over the planning period $[0, \infty[$.

Having completed the description of our durable goods oligopoly game, we now turn to the specification of the equilibrium concept. Throughout the subsequent analysis, we assume Nash-behavior in the sense that firm i takes as given the behavior of its competitors and, in doing so, determines its best response to this behavior. Hence, the equilibrium in our durable good game may be defined as in

Definition D2. $[s_1^*, \ldots, s_n^*]$ *is a rental equilibrium in open-loop strategies (or, for short, open-loop rental equilibrium) if*

(i) all firms face an open-loop information pattern and

(ii) $\Pi^{ri}(s_i^*, s_{-i}^*) \geq \Pi^{ri}(s_i, s_{-i}^*)$ *for all* $s_i \in S$ *and all* $i \in N$.

According to this definition, a rental equilibrium in open-loop strategies is attained if all firms use open-loop strategies and if all firms mutually give the best response to the strategies chosen by the respective competitors. In the subsequent analysis, we denote this equilibrium also as industry equilibrium since it describes the equilibrium in a durable good industry or as market equilibrium since firms take into account the demand function and thus supply equals demand for the services of the durable good.

Determining the Market Equilibrium

The next step is to determine an open-loop rental equilibrium. According to definition D2, this requires deriving the solutions to the firms' profit maximization problems under the assumptions of best-response behavior and open-loop information pattern.

Let us first consider a particular firm i. This firm maximizes its rental profits $\Pi^{ri}(s_i, s_{-i})$ with respect to s_i taking as given the strategies s_{-i} of its competitors. With the help of (2.12)–(2.14), (2.20) and the definitions of s_i and s_{-i}, this maximization problem of firm i may be written as

$$\left. \begin{aligned} & \max_{\substack{y_i(0,\infty) \\ \phi_i(0,\infty)}} \int_0^\infty \left[c_i(t) P[c_i(t) + C^i(t)] - K[y_i(t), \phi_i(t)] \right] e^{-\delta t} dt \\ & \text{subject to} \qquad c_i(t) = \int_{-\infty}^t D[t - v, \phi_i(v)] y_i(v) dv \,, \end{aligned} \right\} \quad (2.21)$$

where $C^i(t) := \sum_{j \in N, \, j \neq i} c_j(t)$ is the durable stock of firm i's competitors. $C^i(t)$ is determined by all firms $j \neq i$ and, hence, firm i takes $C^i(t)$ as given. Furthermore, under the open-loop assumption, firm i irrevocably determines the time paths of output and durability right at the beginning of the planning period. (2.21) is then an optimal control problem where $y_i(t)$ and $\phi_i(t)$ are the control variables and $c_i(t)$ represents the state variable.

Necessary conditions for the solution of optimal control problems are usually derived with the help of the so-called maximum principle. For the 'standard' control problem comprising ordinary differential equation for the state variables, the maximum principle is derived by Pontryagin et al. (1962). But the state variable in (2.21) is not determined by an ordinary differential equation, but by an integral equation.[4] The maximum principle for optimal control problems with integral state equations is provided by Bakke (1974, Theorem 2.1). Applying Bakke's general result to our durable good model requires defining the present-value Hamiltonian as

$$\mathcal{H}[y_i(t), \phi_i(t), c_i(t), \lambda_i(t, \infty), t] = \Big[c_i(t)P[C(t)] - K[y_i(t), \phi_i(t)]\Big]e^{-\delta t}$$

$$+ \int_t^{\infty} D\big[v - t, \phi_i(t)\big]y_i(t)\lambda_i(v)dv \, , \quad (2.22)$$

where $\lambda_i(t)$ is the costate variable associated with the state variable $c_i(t)$. The Hamiltonian in (2.22) is a function $\mathcal{H} : \mathbb{R}^3_+ \times M(t, \infty) \times [0, \infty[\mapsto \mathbb{R}$ where $M(a, b) := \{x(a, b)|a, b \in \mathbb{R}\}$. It measures the impact of time t control and state variables on the objective function in the maximization problem (2.21), i.e. on firm i's rental profits. Changes in the state variable $c_i(t)$ exert a single effect on these profits. They directly alter rental revenues at time t. This effect is captured by the term $c_i(t)P[C(t)]$ in the Hamiltonian (2.22). In contrast, variations of time t control variables $y_i(t)$ and $\phi_i(t)$ have a direct and an indirect impact on the objective function. On the one hand, they directly influence rental profits through changes in production costs at time t. This effect is reflected by the cost function K in the Hamiltonian (2.22). On the other hand, changes in $y_i(t)$ and $\phi_i(t)$ have an indirect effect on the objective function since they alter the durable stock at times $v \geq t$, i.e. at time t and in some non-empty time interval immediately following t.[5] This effect is captured by the integral term in the Hamiltonian (2.22) where the costate variable $\lambda_i(v)$ can be interpreted as shadow price of firm i's stock of the durable good at time v.

Applying the maximum principle of Bakke (1974) to firm i's profit maximization problem (2.21) immediately yields

Proposition 2.1. Let $\big[y_i^*(0, \infty), \phi_i^*(0, \infty)\big]$ be the solution to (2.21) and let $c_i^*(0, \infty)$ be the associated path of the state variable satisfying the integral state constraint in (2.21). Then there exists a path $\lambda_i(0, \infty)$ such that

---

[4] By differentiating the integral state equation with respect to time, we may transform problem (2.21) into a control problem with a differential equation. But the resulting differential equation is not an ordinary one since it still contains an integral. Consequently, the maximum principle of Pontryagin et al. (1962) can not be applied to problem (2.21).

[5] This time interval is compact if the lifetime of the goods is finite, for example, in case of one-hoss shay products. If the decay function is exponential, however, then changes in $y_i(t)$ and $\phi_i(t)$ change the stock of the durable good at *all* times v in the open interval $[t, \infty[$. In order to avoid complicated phrases, throughout the subsequent analysis we simply say that changes in output and durability alter the stock of the durable good 'at times $v \geq t$'.

$$[y_i^*(t), \phi_i^*(t)] = \arg \max_{y_i(t), \phi_i(t)} \mathcal{H}[y_i(t), \phi_i(t), c_i^*(t), \lambda_i(t, \infty), t] , \quad (2.23)$$

$$\lambda_i(t) = \mathcal{H}_c[y_i^*(t), \phi_i^*(t), c_i^*(t), \lambda_i(t, \infty), t]$$

$$= \Big[P[C(t)] + c_i^*(t) P'[C(t)] \Big] e^{-\delta t} , \quad (2.24)$$

for all $t \in [0, \infty[$.

Proposition 2.1 characterizes the solution to firm i's profit maximization problem (2.21). According to condition (2.24), the shadow price of firm i's stock of the durable good at time t equals marginal revenues of this stock at time t. Condition (2.23) states that at every time t of the planning period, firm i has to set the control variables such that, for an optimal stock of the durable good, the Hamiltonian at time t attains a maximum, i.e. such that the total (direct and indirect) impact of time t output and durability on firm i's rental profits is maximized.

Proposition 2.1 provides a rather general characterization of the solution to (2.21) since condition (2.23) allows for corner solutions for output and product durability. Such corner solutions will play an important role in Chap. 3 when we analyze production emissions of durable goods. For the time being, however, we follow Sieper and Swan (1973), Goering (1993) and Goering and Boyce (1999) and suppose interior solutions for $y_i(t)$ and $\phi_i(t)$. Condition (2.23) then implies[6]

$$\mathcal{H}_y(t) = -K_y(t)e^{-\delta t} + \int_t^\infty D(v)\lambda_i(v)dv = 0 , \quad (2.25)$$

$$\mathcal{H}_\phi(t) = -K_\phi(t)e^{-\delta t} + y_i^*(t) \int_t^\infty D_\phi(v)\lambda_i(v)dv = 0 , \quad (2.26)$$

for all $t \in [0, \infty[$. Inserting (2.24) into these conditions yields

$$K_y(t) = \int_t^\infty D(v)[P(v) + c_i^*(v)P'(v)]e^{-\delta(v-t)}dv , \quad (2.27)$$

$$\frac{K_\phi(t)}{y_i^*(t)} = \int_t^\infty D_\phi(v)[P(v) + c_i^*(v)P'(v)]e^{-\delta(v-t)}dv , \quad (2.28)$$

for all $t \in [0, \infty[$. (2.27) and (2.28) possess the usual 'marginal-cost-marginal-revenues' interpretation. The LHS of (2.27) equals marginal production costs of firm i's time t output. The RHS of (2.27) represents marginal revenues of

---

[6] For notational convenience, if there is no risk of confusion, we mark functions only with their time index t or v and drop all other arguments, e.g. $K_y(t) := K_y[y_i^*(t), \phi_i^*(t)]$. The decay function and its derivatives are abbreviated by $D(v) := D[v - t, \phi_i^*(t)]$, $D_a(v) := D_a[v - t, \phi_i^*(t)]$ and so on.

firm i's time t output. Note that an increase in time t output increases firm i's stock of the durable good not only at time t, but also at times $v > t$ since the good under consideration is durable. Therefore, marginal revenues of time t output reflect marginal revenues of the durable stock accruing in the interval $[t, \infty[$. In sum, equation (2.27) equates marginal costs and marginal revenues of firm i's time t output. Equation (2.28) states that marginal costs of firm i's time t durability have to equal marginal revenues. The LHS of (2.28) reflects marginal production costs of time t durability per unit of output. The RHS of (2.28) equals the pertaining marginal revenues. These marginal revenues arise at times $v \geq t$ since an increase in time t product durability raises firm i's stock of the durable good not only at t, but also at times $v > t$.

It should be noted that (2.27) and (2.28) together with the integral constraint in (2.21) are necessary for the solution of firm i's profit maximization problem since Proposition 2.1 provides necessary conditions only. As shown by Kamien and Muller (1976, Theorem 2), however, these conditions are also sufficient if the maximized Hamiltonian

$$\mathcal{H}^{\max}[c_i(t), \lambda_i(t, \infty), t] := \max_{y_i(t), \phi_i(t)} \mathcal{H}[y_i(t), \phi_i(t), c_i(t), \lambda_i(t, \infty), t] \quad (2.29)$$

is concave in the state variable $c_i(t)$ for any given $\lambda_i(t, \infty)$ and all $t \in [0, \infty[$.[7] It is straightforward to show that this concavity condition is satisfied for the maximization problem at hand: For given $c_i(t)$ and $\lambda_i(t, \infty)$, the maximum of $\mathcal{H}$ with respect to $y_i(t)$ and $\phi_i(t)$ is determined by (2.25) and (2.26). These conditions do not contain the durable stock $c_i(t)$ and, hence, the solutions for $y_i(t)$ and $\phi_i(t)$ are independent of the durable stock. The concavity of the maximized Hamiltonian (2.29) is then proved by differentiating the Hamiltonian (2.22) with respect to $c_i(t)$ taking as given $y_i(t)$ and $\phi_i(t)$. According to assumption A4, this yields $\mathcal{H}_{cc}^{\max}(t) = [2P'(t) + c_i(t)P''(t)]e^{-\delta t} < 0$ for all $t \in [0, \infty[$. Hence, the maximized Hamiltonian is concave in the durable stock and (2.27) and (2.28) together with the integral constraint for $c_i(t)$ in (2.21) are necessary *and* sufficient for the solution of firm i's profit maximization problem.

As already argued above, a market equilibrium of the renting durable good oligopoly is attained if all firms mutually give the best response to the behavior of the respective competitors. Hence, every firm maximizes rental profits with respect to its choice variables taking as given the choice variables of its rivals. The equilibrium time paths of the firms' output, durability and stock of the durable good are therefore determined by (2.27), (2.28) and the integral constraint in (2.21) for all $t \in [0, \infty[$ and all $i \in N$. The major part of the subsequent analysis focuses on a symmetric equilibrium with the equilibrium time paths $y_i^*(0, \infty) = y^*(0, \infty)$, $\phi_i^*(0, \infty) = \phi^*(0, \infty)$ and $c_i^*(0, \infty) = c^*(0, \infty)$ for all $i \in N$. The equilibrium conditions then read

---

[7] This sufficiency theorem is of the Arrow-type as defined in Seierstad and Sydæster (1987, p. 107).

$$K_y[y^*(t), \phi^*(t)] = \int_t^\infty D[v - t, \phi^*(t)] \times$$

$$\times \Big[P[nc^*(v)] + c^*(v)P'[nc^*(v)] \Big] e^{-\delta(v-t)} dv , \quad (2.30)$$

$$\frac{K_\phi[y^*(t), \phi^*(t)]}{y^*(t)} = \int_t^\infty D_\phi[v - t, \phi^*(t)] \times$$

$$\times \Big[P[nc^*(v)] + c^*(v)P'[nc^*(v)] \Big] e^{-\delta(v-t)} dv , \quad (2.31)$$

$$c^*(t) = \int_{-\infty}^t D[t - v, \phi^*(v)]y^*(v)dv , \quad (2.32)$$

for $t \in [0, \infty[$. This is a set of three integral equations determining the equilibrium time paths $y^*(0, \infty)$, $\phi^*(0, \infty)$ and $c^*(0, \infty)$. For explicitly solving (2.30)–(2.32) with respect to the equilibrium time paths, we actually need the transversality condition for the costate variable $\lambda_i(t)$ and initial conditions $y_i(-\infty, 0) = y_i^0(-\infty, 0)$ and $\phi_i(-\infty, 0) = \phi_i^0(-\infty, 0)$ where $y_i^0(-\infty, 0)$ and $\phi_i^0(-\infty, 0)$ are exogenously given time paths, $i \in N$. However, the subsequent analysis is restricted to a qualitative characterization of the market equilibrium and, therefore, transversality and initial conditions are ignored throughout.

Sales versus Rental, Open-Loop versus Closed-Loop Equilibrium

The market equilibrium conditions (2.30)–(2.32) are derived under the assumptions that firms rent their output and employ open-loop strategies. We will now discuss whether the equilibrium changes if these assumptions are removed.

Let us first retain the open-loop assumption and consider the case in which firms sell their products. In order to characterize the pertinent equilibrium, two alternative assumptions are introduced.

Assumption A5. *In both the rental and the sales case, all firms start production not before time zero so that $y_i(v) = \phi_i(v) = 0$ for all $v \in\,]-\infty, 0[$ and all $i \in N$.*

Assumption A6. *In both the rental and the sales case, all firms start production before time zero and, at time $t = 0$, not only renting firms possess the remaining units of vintages produced prior to the planning period, but also do selling firms.*

For deriving the pay-off functions of selling firms, note that the consumers' willingness-to-pay for a unit of the durable good equals the present value of the consumers' willingness-to-pay for the services generated by the unit during its whole life.[8] Consequently, the sales price of a unit of the durable good equals the present value of the associated rentals. Under assumption A5, the selling firm sells the vintages produced during the planning period $[0, \infty[$. The sales price of a unit of firm i's vintage $v \in [0, \infty[$ equals $\int_v^\infty D[t - v, \phi_i(v)]P[C(t)]e^{-\delta(t-v)} \, dt$. The pay-off of the selling firm i under assumption A5 is therefore a function $\Pi^{si} : \mathcal{S}^n \mapsto \mathbb{R}$ where

$$\Pi^{si}(s_i, s_{-i}) = \int_0^\infty \Bigg[y_i(v) \int_v^\infty D[t - v, \phi_i(v)] \times$$

$$\times P\Bigg[\sum_{j \in N} \int_{-\infty}^t D[t - \tilde{v}, \phi_j(\tilde{v})] y_j(\tilde{v}) d\tilde{v} \Bigg] e^{-\delta(t-v)} \, dt$$

$$- K[y_i(v), \phi_i(v)] \Bigg] e^{-\delta v} \, dv \quad (2.33)$$

is the present value of firm i's profits from selling the vintages $v \in [0, \infty[$. Under assumption A6, the selling firm sells not only the units produced during the planning period, but also the remaining units of vintages produced prior to time zero. Buying these remaining units at time zero, consumers are willing to pay only for the services which the units provide during their remaining life. Hence, at time zero, the sales price of a unit of firm i's vintage $v \in]-\infty, 0[$ is $\int_0^\infty D[t - v, \phi_i(v)]P[C(t)]e^{-\delta(t-v)} \, dt$. Note that we integrate over $[0, \infty[$ and not over $[v, \infty[$. The revenues from selling the remaining units of vintages $v \in]-\infty, 0[$ at $t = 0$ are therefore

$$\Pi_o^{si}(s_i, s_{-i}) = \int_{-\infty}^0 \Bigg[y_i(v) \int_0^\infty D[t - v, \phi_i(v)] \times$$

$$\times P\Bigg[\sum_{j \in N} \int_{-\infty}^t D[t - \tilde{v}, \phi_j(\tilde{v})] y_j(\tilde{v}) d\tilde{v} \Bigg] e^{-\delta(t-v)} \, dt \Bigg] e^{-\delta v} \, dv \,. \quad (2.34)$$

It follows that, under assumption A6, the pay-off of the selling firm i is a function $\widetilde{\Pi}^{si} : \mathcal{S}^n \mapsto \mathbb{R}$ where

$$\widetilde{\Pi}^{si}(s_i, s_{-i}) = \Pi^{si}(s_i, s_{-i}) + \Pi_o^{si}(s_i, s_{-i}) \quad (2.35)$$

is the present value of firm i's profits from selling the vintages produced during the planning period and from selling the remaining units of vintages produced prior to the planning period.

---

[8] This will become clear in Chap. 5 which presents general equilibrium models accounting for a (representative) utility-maximizing durable good consumer. See especially footnote 7 on p. 153.

The definition of an open-loop sales equilibrium is almost the same as the definition of an open-loop rental equilibrium presented in definition D2. The only difference is that the rental profits $\Pi^{ri}(s_i, s_{-i})$ have to be replaced by the sales profits $\Pi^{si}(s_i, s_{-i})$ and $\widetilde{\Pi}^{si}(s_i, s_{-i})$, respectively. We then obtain

Proposition 2.2. *Suppose either assumption A5 or assumption A6 is satisfied. Then, the open-loop sales equilibrium is identical to the open-loop rental equilibrium.*

Proof. The idea of the proof is to show that, under assumption A5 (A6), the pay-off function Π^{si} ($\widetilde{\Pi}^{si}$) is identical to the pay-off function Π^{ri} in the rental case. The derivation of the open-loop sales equilibrium is then exactly the same as the derivation of the open-loop rental equilibrium presented in the previous paragraph of this section.

Let us start with markets where assumption A5 is satisfied. Firm i's pay-off function in (2.33) may be written as

$$\Pi^{si}(s_i, s_{-i}) = \iint\limits_{S_1} D[t - v, \phi_i(v)]y_i(v)P[\cdot]e^{-\delta t}dv\,dt$$

$$- \int_0^\infty K[y_i(t), \phi_i(t)]e^{-\delta t}dt\,,$$

where $S_1 := \{(v,t)|\, v \in [0, \infty[\, \wedge\, t \in [v, \infty[\}$ is the support of the integration variables v and t. Moreover, assumption A5 implies that the pay-off function (2.20) in the rental case becomes

$$\Pi^{ri}(s_i, s_{-i}) = \iint\limits_{S_2} D[t - v, \phi_i(v)]y_i(v)P[\cdot]e^{-\delta t}dv\,dt$$

$$- \int_0^\infty K[y_i(t), \phi_i(t)]e^{-\delta t}dt\,,$$

with $S_2 := \{(v,t)|\, v \in [0, t]\, \wedge\, t \in [0, \infty[\}$. It is straightforward to show that in the (v, t)-plane, we obtain $S_1 = S_2$. Hence, $\Pi^{si}(s_i, s_{-i}) = \Pi^{ri}(s_i, s_{-i})$ and the open-loop sales equilibrium is the same as the open-loop rental equilibrium.

The analogous argument applies in case that assumption A6 holds. The pay-off function (2.35) of the selling firm may be written as

$$\widetilde{\Pi}^{si}(s_i, s_{-i}) = \iint\limits_{S_1 \cup S_3} D[t - v, \phi_i(v)]y_i(v)P[\cdot]e^{-\delta t}dv\,dt$$

$$- \int_0^\infty K[y_i(t), \phi_i(t)]e^{-\delta t}dt\,,$$

with $S_3 := \{(v,t)\,|\,v \in]-\infty,0] \ \wedge\ t \in [0,\infty[\}$ whereas the pay-off function
(2.20) of the renting firm reads

$$\Pi^{ri}(s_i,s_{-i}) = \iint\limits_{S_4} D[t - v, \phi_i(v)]y_i(v)P[\cdot]e^{-\delta t}dv\,dt$$

$$- \int_0^\infty K[y_i(t),\phi_i(t)]e^{-\delta t}dt\,,$$

where $S_4 := \{(v,t)\,|\,v \in]-\infty,t] \ \wedge\ t \in [0,\infty[\}$. In the (v,t)-plane, we obtain
$S_1 \cup S_3 = S_4$. This implies $\widetilde{\Pi}^{si}(s_i,s_{-i}) = \Pi^{ri}(s_i,s_{-i})$ with the consequence
that open-loop sales and rental equilibria coincide.   □

According to Proposition 2.2, conditions (2.30)–(2.32) determine not only a
rental equilibrium, but also a sales equilibrium provided firms employ open-
loop strategies and one of the assumptions A5 and A6 is satisfied. To judge
the relevance of this result, we present examples in which the underlying assumptions hold. Assumption A5 is satisfied for all newly introduced durable
goods. Every new generation of mobile phones or personal computers are recent examples. Finding an example for assumption A6 is more difficult since in
the sales case the property of the products usually passes over to consumers.
Nevertheless, assumption A6 is yet satisfied if sellers and buyers sign leasing
contracts instead of trading the durable directly. The units produced prior to
time zero are then in property of consumers', indeed, but the revenues from
this products accrue to the producers not only before, but also during the planning period. This is equivalent to selling the remaining units at the beginning
of the planning period. Another (more artificial) example for assumption A6
is the case in which firms switch from a rental policy to a sales policy at the
beginning of the planning period and sell all the remaining units of vintages
produced before the planning period. The assumption of an open-loop information pattern is satisfied whenever firms possess commitment ability in the
sense that they can credibly precommitment to their time paths at time zero
already. Mechanisms which endow durable good sellers with such commitment
ability are, for example, binding leasing contracts between sellers and buyers,
potential entry of competitors (Ausubel and Deneckere, 1987), reputation of
sellers (Ausubel and Deneckere, 1989) or best-price provision (Butz, 1990).

In many cases, however, producers employ closed-loop strategies instead
of open-loop ones since they do not possess commitment ability. We will now
discuss the consequences of the closed-loop assumption with respect to the
rental and sales equilibria. Let us start with the rental case. In a two-period
model without environmental pollution, Goering (1992) shows that the closed-
loop rental equilibrium is identical to the open-loop rental equilibrium. We
will derive the same result in Sect. 4.2 where a two-period durable good model
with solid consumption waste is investigated. The rationale for the equivalence
of open-loop and closed-loop rental equilibria is that renting firms remain the

owner of the durable stock and, hence, it is in the firm's own interest to account for the future value of the durable already today. Consequently, renting firms do not receive incentives to change their initial plan in future periods even if they possess no commitment ability. Unfortunately, it has to be left open whether this result is also true in the multi-period vintage approach considered in the present section since, to the best of my knowledge, general solutions for a closed-loop equilibrium of such a multi-period game have not yet been developed in the literature.

Now turn to the sales case without precommitment ability of producers. In a two-period model without pollution, Bulow (1986) shows that the closed-loop sales equilibrium differs from the open-loop sales equilibrium, in general. Again, the same result is obtained in the two-period durable good model with solid waste considered in Sect. 4.2. The rationale for this difference between open-loop and closed-loop sales equilibria lies in the fact that durable goods sellers are not the owner of the durable stock. Consequently, in future periods, a selling firm without precommitment ability has an incentive to deviate from its initially chosen plan: It supplies more units of the durable good tomorrow than it announces today since the loss in value (the reduction in the market price due to the additional supply) is borne by the owners of the durable stock which in case of sales markets are consumers. Unfortunately, we are again not able to prove this result in the vintage model of the present section since solution procedures for a closed-loop equilibrium are not available. But it seems to be plausible that open-loop and closed-loop sales equilibria differ also in a multi-period approach.

2.2.2 Stationary State

In dynamic models, one usually distinguish between the (short run) adjustment path on which the model variables are subject to changes over time and the (long run) stationary state[9] in which all model variables are constant over time. In this section, we will present and discuss alternative definitions of the stationary state and then characterize the stationary state of the market equilibrium derived in the previous section. For later reference, also the firm's long run profit maximization problem will be considered.

Definition of the Stationary State and Long Run Market Equilibrium

Let us start with the definition employed in previous durable good studies. Sieper and Swan (1973) consider the stationary state as a '... situation in

---

[9] In the subsequent analysis, the terms 'stationary state', 'steady state' and 'long run' are used as synonyms. The same is done with the terms 'adjustment path' and 'short run' although the adjustment path may prevail forever, i.e. in the long run. These conventions will help to avoid clumsy phrases.

which durability, the production rate and the aggregated service flow are all constant over time.' (Sieper and Swan, 1973, p. 335). In fact, they use

Definition D3. (y, ϕ, c) *is called stationary state if* $y(t) = y$, $\phi(t) = \phi$ *and* $c(t) = c$ *for all* $t \in \mathbb{R}$.

The same definition is used by Abel (1983), Goering (1993) and Goering and Boyce (1999). If we suppose that the open-loop market equilibrium derived in the previous section attains a stationary state as defined in definition D3, then we have $y^*(t) = y^*$, $\phi^*(t) = \phi^*$ and $c^*(t) = c^*$ for all $t \in \mathbb{R}$ and the market equilibrium conditions (2.30)–(2.32) become

$$K_y(y^*, \phi^*) = [P(nc^*) + c^* P'(nc^*)] \Psi(\phi^*) \,, \tag{2.36}$$

$$\frac{K_\phi(y^*, \phi^*)}{y^*} = [P(nc^*) + c^* P'(nc^*)] \Psi'(\phi^*) \,, \tag{2.37}$$

$$c^* = y^* \Gamma(\phi^*) \,, \tag{2.38}$$

where we have used the transformation $a := t - v$ and where Γ and Ψ are the durable's undiscounted and discounted use potential introduced in Sect. 2.1.2. Equations (2.36)–(2.38) characterize the long run market equilibrium when definition D3 is applied. They determine the long run equilibrium values y^*, ϕ^* and c^*.

Definition D3 requires output, durability and the stock of the durable good to be constant over the whole time axis, i.e. for all $t \in \mathbb{R}$. This implies (a) that there is no adjustment path on which output, durability and the stock of the durable good are subject to changes over time and (b) that prior to the planning period incidentally the same values of output, durability and stock of the durable good prevail which producers choose during the planning period. Sieper and Swan (1973, pp. 338) identify conditions under which point (a) is satisfied, but it remains unclear whether point (b) is also satisfied. They simply observe that, before the planning period, output and durability '... must have been constant at the levels which are currently optimal if long run equilibrium is to persist ...' (Sieper and Swan, 1973, p. 345), but they do not give an explanation why or when this should be the case. Nevertheless, we will now introduce two alternative steady state definitions which require neither condition (a) nor condition (b) and then show that, under some qualifications, the long run market equilibrium conditions derived on the basis of these alternative definitions are exactly the same as (2.36)–(2.38).

The first alternative to definition D3 is

Definition D4. (y, ϕ, c) *is called stationary state if* $y(t) = y$, $\phi(t) = \phi$ *and* $c(t) = c$ *for all* $t \geq t_s$ *with* $t_s \in \mathbb{R}_+$.

In contrast to definition D3, definition D4 allows for an adjustment path in the interval $] - \infty, t_s[$ on which output, durability and the stock of the durable good may take different values for different periods in time. Furthermore, definition D4 does not require that prior to the planning period the model variables take on the same values as during the planning period since t_s is non-negative. To characterize the long run market equilibrium on the basis of definition D4, we introduce

Assumption A7. *There exists a finite and non-negative number $\bar{t}$ such that $y^*(t) = y^*$ and $\phi^*(t) = \phi^*$ for all $t \in [\bar{t}, \infty[$.*

This assumption says that equilibrium output and durability attain constant values in finite time. The equilibrium stock of the durable good in (2.32) may then be written as

$$c^*(t) = \underbrace{\int_{-\infty}^{\bar{t}} D[t - v, \phi^*(v)] y^*(v) dv}_{=:I(t)} + y^* \int_{\bar{t}}^{t} D(t - v, \phi^*) dv , \quad (2.39)$$

for all $t \in [\bar{t}, \infty[$. This equation shows that constant output and durability after time $\bar{t}$ do not yet imply a constant stock of the durable good after time $\bar{t}$. $c^*(t)$ in (2.39) is influenced also by the vintages produced prior to $\bar{t}$ which are reflected by the time-dependent term $I(t)$. To ensure that the stock of the durable good is constant, too, we introduce an additional assumption on the decay function.

Assumption A8. *The decay function (2.1) satisfies the condition:*

$\forall \phi \in \mathbb{R}_+ \; \exists$ *a finite positive number $A(\phi)$ such that $D(a, \phi) = 0 \; \forall a \geq A(\phi)$.*

This assumption is more restrictive than condition (2.2) in assumption A1. Roughly speaking, it states that every unit of the durable good wears out in finite time for all feasible values of durability. Due to assumption A8 there exists $t_s > \bar{t}$ such that $I(t) = 0$ for all $t \in [t_s, \infty[$. Hence, from (2.39) it follows

$$c^*(t) = y^* \int_{\bar{t}}^{t} D(t - v, \phi^*) dv , \quad (2.40)$$

for all $t \in [t_s, \infty[$. Let $T(\phi^*)$ be the age of the oldest vintage in use at time $t \geq t_s$. $T(\phi^*)$ is the same for all $t \geq t_s$ since equilibrium durability is the same for all $t \geq t_s$. The stock of the durable good in (2.40) can then be written as

$$c^*(t) = y^* \int_{t-T(\phi^*)}^{t} D(t - v, \phi^*) dv$$

$$= y^* \underbrace{\int_{0}^{T(\phi)} D(a, \phi^*) da}_{=:\tilde{\Gamma}(\phi^*)} = y^* \tilde{\Gamma}(\phi^*) =: c^* , \quad (2.41)$$

for all $t \in [t_s, \infty[$. Consequently, (y^*, ϕ^*, c^*) is a stationary state as defined in definition D4. Due to assumption A8 we obtain $\tilde{\Gamma}(\phi^*) = \Gamma(\phi^*)$ and (2.41) becomes

$$c^* = y^* \Gamma(\phi^*) , \qquad (2.42)$$

for all $t \in [t_s, \infty[$. Setting $y^*(t) = y^*$, $\phi^*(t) = \phi^*$ and $c^*(t) = c^*$ for $t \in [t_s, \infty[$ in the remaining market equilibrium conditions (2.30) and (2.31) yields

$$K_y(y^*, \phi^*) = [P(nc^*) + c^* P'(nc^*)] \Psi(\phi^*) , \qquad (2.43)$$

$$\frac{K_\phi(y^*, \phi^*)}{y^*} = [P(nc^*) + c^* P'(nc^*)] \Psi'(\phi^*) , \qquad (2.44)$$

for all $t \in [t_s, \infty[$. Hence, when we presuppose definition D4 and introduce the assumptions A7 and A8, then (2.42)–(2.44) are the long run market equilibrium conditions. As already mentioned above, these conditions are identical to the long run market equilibrium conditions (2.36)–(2.38) derived by Sieper and Swan (1973) using definition D3.

In deriving (2.42)–(2.44), we have used assumptions A7 and A8. Assumption A8 is clearly realistic since owing to the law of entropy no durable good has an infinite lifetime; all durables are bound to be scrapped in finite time. One-hoss shay durable goods are an example having this property. In our formal analysis, however, we also have introduced in Sect. 2.1.2 the (generalized) exponential decay function which satisfies not assumption A8, but the less restrictive condition (2.2) in assumption A1. Furthermore, also assumption A7 may be too restrictive for our formal analysis since it is possible that equilibrium output and durability do not attain constant values in finite time, but satisfy the less restrictive conditions in[10]

Assumption A9. $\lim_{t \to \infty} y^*(t) = y^*$ *and* $\lim_{t \to \infty} \phi^*(t) = \phi^*$.

If our model does not satisfy the assumptions A7 and A8, but the less restrictive assumption A9 and condition (2.2) in assumption A1, then the market equilibrium does not reach a stationary state as defined in definition D4. However, we will now introduce a third steady state definition and show that the market equilibrium attains a stationary state in the sense of this definition if assumption A9 and condition (2.2) in assumption A1 are satisfied. It will also turn out that the associated long run equilibrium conditions are the same as (2.42)–(2.44).

The third steady state definition is

Definition D5. (y, ϕ, c) *is called asymptotic stationary state if*

$$\lim_{t \to \infty} y(t) = y , \qquad \lim_{t \to \infty} \phi(t) = \phi \qquad and \qquad \lim_{t \to \infty} c(t) = c .$$

---

[10] The solution to dynamic models often takes the form of exponential functions, for example, if the model is described by linear differential equations. Provided the pertinent stability conditions are satisfied, the model variables then asymptotically converge to constant levels, but never reach these levels. See e.g. the textbook of Shone (1998).

As definition D4, also definition D5 allows for an adjustment path and for different values of the model variables before and during the planning period. In contrast to definition D4, however, definition D5 does not require that the model variables attain constant values in finite time. Output, durability and stock of the durable good only have to converge to constant levels. To characterize the long run market equilibrium on the basis of definition D5, suppose the market equilibrium satisfies assumption A9 and let $\hat{t} \in \mathbb{R}_+$ a point in time sufficiently large such that equilibrium output and equilibrium durability are already close to their constant levels y^* and ϕ^*, respectively. From the market equilibrium condition (2.32) we then obtain

$$\lim_{t \to \infty} c^*(t) = \lim_{t \to \infty} \underbrace{\int_{-\infty}^{\hat{t}} D[t - v, \phi^*(v)]y^*(v)dv}_{=:\widehat{I}(t)}$$

$$+ y^* \lim_{t \to \infty} \int_{\hat{t}}^{t} D(t - v, \phi^*)dv . \quad (2.45)$$

Condition (2.2) in assumption A1 imply $\widehat{I}(t) \to 0$ as $t \to \infty$. Using the transformation $a := t - v$, (2.45) becomes

$$\lim_{t \to \infty} c^*(t) = y^* \lim_{t \to \infty} \int_{0}^{t - \hat{t}} D(a, \phi^*)da$$

$$= y^* \int_{0}^{\infty} D(a, \phi^*)da = y^* \Gamma(\phi^*) =: c^* . \quad (2.46)$$

Hence, (y^*, ϕ^*, c^*) is an asymptotic stationary state as defined in definition D5. Taking into account (2.46) and assumption A9, the remaining market equilibrium conditions (2.30) and (2.31) converge to

$$K_y(y^*, \phi^*) = [P(nc^*) + c^* P'(nc^*)]\Psi(\phi^*) , \quad (2.47)$$

$$\frac{K_\phi(y^*, \phi^*)}{y^*} = [P(nc^*) + c^* P'(nc^*)]\Psi'(\phi^*) . \quad (2.48)$$

Consequently, when we presuppose definition D5 and account for assumption A9 and condition (2.2) in assumption A1, then (2.46)–(2.48) are the conditions determining the asymptotic long run market equilibrium (y^*, ϕ^*, c^*).[11] As mentioned above, these conditions are the same as (2.36)–(2.38) and (2.42)–(2.44) which represent the long run market equilibrium conditions derived on the basis of definition D3 and D4, respectively.

---

[11] Of course, this result also holds if we combine either assumption A9 with assumption A8 or condition (2.2) in assumption A1 with assumption A7.

In sum, definition D5 is less restrictive than the other steady state definitions and, under some qualifications, leads to the same long run equilibrium conditions as these definitions. The subsequent analysis therefore presupposes definition D5 together with assumption A9 and condition (2.2) in assumption A1. This implies that (2.46)–(2.48) determine the asymptotic long run market equilibrium (y^*, ϕ^*, c^*). To avoid clumsy phrases, the term 'asymptotic' is suppressed throughout.

Finally, we briefly discuss existence and stability of the long run market equilibrium. A stationary state of the market equilibrium exists if the set of equations (2.46)–(2.48) possesses a solution with respect to (y^*, ϕ^*, c^*). This can be proved by the way of example: Suppose the demand and cost functions are $P(C) = \beta_1 - \beta_2 C$ and $K(y, \phi) = y^2/(1 - \phi)$, respectively, and the decay function is exponential. Setting $\beta_1 = 100$, $\beta_2 = 0.1$, $n = 2$ and $\delta = 0.04$ and solving (2.46)–(2.48) shows that there exists a long run market equilibrium with $y^* \approx 10.6866$, $\phi^* \approx 0.6608$ and $c^* \approx 7.0621$. Investigating stability requires specifying the conditions under which the equilibrium time paths $y^*(0, \infty)$, $\phi^*(0, \infty)$ and $c^*(0, \infty)$ determined by the integral equations (2.30)–(2.32) converge to the long run equilibrium (y^*, ϕ^*, c^*) determined by (2.46)–(2.48). To the best of my knowledge, however, general stability conditions for a system of integral equations are not yet available in literature and, hence, it is left open whether the long run market equilibrium is stable.

Long Run Profit Maximization

In contrast to the market equilibrium conditions (2.30)–(2.32), the long run market equilibrium conditions (2.46)–(2.48) are not integral equations, but 'ordinary' (static) equations. It is therefore obvious to ask whether (2.46)–(2.48) can be derived with the help of a static maximization problem instead of the dynamic profit maximization problem (2.21). We will now discuss two alternative static maximization problems and investigate which of these problems is appropriate to derive the long run equilibrium conditions (2.46)–(2.48).

For the first maximization problem, suppose the durable good economy has already attained a stationary state. Firm i's long run rental profits read $c_i P(C) - K(y_i, \phi_i)$. If we assume that firm i maximizes long run profits taking into account the long run stock of the durable good, then we obtain the maximization problem

$$\max_{y_i, \phi_i} c_i P(c_i + C^i) - K(y_i, \phi_i) \qquad \text{s.t.} \qquad c_i = y_i \Gamma(\phi_i) \,. \tag{2.49}$$

Employing Lagrangean techniques to solve this problem and considering the symmetric case yields

$$K_y(y, \phi) = [P(nc) + cP'(nc)]\Gamma(\phi) \,, \tag{2.50}$$

$$\frac{K_\phi(y, \phi)}{y} = [P(nc) + cP'(nc)]\Gamma'(\phi) \,, \tag{2.51}$$

$$c = y\Gamma(\phi) . \tag{2.52}$$

In contrast to the long run market equilibrium conditions (2.46)–(2.48), conditions (2.50)–(2.52) do not contain the discounted use potential Ψ and its derivative Ψ', but the undiscounted use potential Γ and its derivative Γ'. It immediately follows

Proposition 2.3. *Let $(\hat{y}, \hat{\phi}, \hat{c})$ be the solution to (2.50)–(2.52). Then $(\hat{y}, \hat{\phi}, \hat{c}) \neq (y^*, \phi^*, c^*)$.*

According to Proposition 2.3, the long run maximization problem (2.49) is not suitable to derive the long run market equilibrium determined by (2.46)–(2.48). The reason for this result is as follows. The long run market equilibrium is derived from a dynamic model which comprises an adjustment path, in general. In contrast, by the constraint $c_i = y_i\Gamma(\phi_i)$ the maximization problem (2.49) supposes that the stock of the durable is on its long run level at any point in time and, hence, (2.49) completely ignores the adjustment path of the durable good economy.

Based on this insight, as a second long run maximization problem, we consider

$$\max_{y_i, \phi_i} c_i P(c_i + C^i) - K(y_i, \phi_i) \text{ s.t. } c_i = y_i\Psi(\phi_i) - \nu_1\Psi(\nu_2) + \nu_1\Gamma(\nu_2) , \tag{2.53}$$

where $\nu_1, \nu_2 \in \mathbb{R}_+$ are given parameters to be specified later. As the long run problem (2.49), also problem (2.53) states that firm i maximizes long run rental profits. In contrast to (2.49), however, (2.53) does not take into account the long run condition for the stock of the durable good, but a modified constraint whose rationale becomes obvious later. Deriving FOCs of problem (2.53) and considering the symmetric case yields

$$K_y(y, \phi) = [P(nc) + cP'(nc)]\Psi(\phi) , \tag{2.54}$$

$$\frac{K_\phi(y, \phi)}{y} = [P(nc) + cP'(nc)]\Psi'(\phi) , \tag{2.55}$$

$$c = y\Psi(\phi) - \nu_1\Psi(\nu_2) + \nu_1\Gamma(\nu_2) . \tag{2.56}$$

The solution of these equations with respect to output, durability and stock of the durable good depends on the parameters ν_1 and ν_2. We then obtain

Proposition 2.4. *Let $[\breve{y}(\nu_1, \nu_2), \breve{\phi}(\nu_1, \nu_2), \breve{c}(\nu_1, \nu_2)]$ be the solution to (2.54)–(2.56). Then*

$$\nu_1 = y^* \wedge \nu_2 = \phi^* \quad \Rightarrow \quad [\breve{y}(\nu_1, \nu_2), \breve{\phi}(\nu_1, \nu_2), \breve{c}(\nu_1, \nu_2)] = (y^*, \phi^*, c^*) .$$

Proof. Inserting $\nu_1 = y^*$ and $\nu_2 = \phi^*$ into (2.56) gives

$$c = y\Psi(\phi) - y^*\Psi(\phi^*) + y^*\Gamma(\phi^*) . \tag{2.57}$$

(y^*, ϕ^*, c^*) is then the solution to (2.54), (2.55) and (2.57). This can be seen by inserting (y^*, ϕ^*, c^*) into (2.54), (2.55) and (2.57) and comparing the resulting expressions with (2.46)–(2.48). $\square$

According to Proposition 2.4, the long run maximization problem (2.53) is suitable to determine the long run market equilibrium provided the exogenous parameters ν_1 and ν_2 are equal to the long run equilibrium values of output and durability, respectively. The reason for this result is that (2.53) accounts for the adjustment path of the durable good economy since it does not suppose the stock of the durable good is always on its long run level $c_i = y_i \Gamma(\phi_i)$.

To sum up, deriving the long run market equilibrium conditions which account for the adjustment path of the durable good economy requires maximizing the firms' long run profits and taking into account the modified constraint $c_i = y_i \Psi(\phi_i) - \nu_1 \Psi(\nu_2) + \nu_1 \Gamma(\nu_2)$ for the long run stock of the durable good. In the resulting FOCs, we have to set the parameter ν_1 and ν_2 equal to the long run equilibrium values of output and durability, respectively.[12]

2.2.3 Comparative Statics of the Long Run Market Equilibrium

According to (2.46)–(2.48), the long run equilibrium values of output, durability and the stock of the durable good are functions of the model parameters as, for example, the number of firms operating in the industry, n, or the common discount rate, δ. Of special interest in the subsequent analysis is how the equilibrium values change if some of the model parameters are varied shifting the economy from one long run equilibrium to another long run equilibrium. This question can be answered with the help of a comparative static analysis[13] which we introduce in the present section.

In previous durable good studies, two different procedures have been suggested to carry out the comparative static analysis of the long run market equilibrium.

Procedure P1. (Abel, 1983) *Totally differentiate (2.46)–(2.48). Apply then Cramer's rule in order to compute the marginal changes in the long*

---

[12] Martin (1962), Kleiman and Ophir (1966), Levhari and Srinivasan (1969) and Schmalensee (1970) use a long run maximization problem like (2.49) to find theoretical arguments in support of the planned obsolescence hypothesis. Sieper and Swan (1973) first point out that such a maximization problem is inappropriate since it ignores the adjustment path to the stationary state. They do not present problem (2.53), but their procedure is equivalent to the one presented here.

[13] It would also be appropriate to use the term 'comparative dynamic analysis' since the long run equilibrium prevails not only for one point in time, but over a whole time interval. Hence, a sensitivity analysis for one (representative) point in time yields results for the whole time interval. In literature, however, the term 'comparative dynamic analysis' is reserved for the sensitivity analysis of the adjustment path (e.g. Feichtinger and Hartl, 1986). We therefore keep using the term 'comparative static analysis'.

run equilibrium values y^*, ϕ^* *and* c^* *due to marginal changes in the model parameters.*

Procedure P2. (Goering and Boyce, 1999) *Suppose the long run equilibrium values are determined by* (2.54)–(2.56). *Totally differentiate this set of equations and, in doing so, take* ν_1 *and* ν_2 *as exogenously given. Apply then Cramer's rule in order to compute the marginal changes in the equilibrium values* y^*, ϕ^* *and* c^* *due to marginal changes in the model parameters.*

Obviously, both procedures differ since they employ different sets of long run equilibrium conditions. It therefore has to be clarified which one is the correct and exact procedure. If we want to know how parameter changes influence the equilibrium values of output, durability and the stock of the durable good, then it is a mathematical fact that we have to totally differentiate the long run market equilibrium conditions (2.46)–(2.48). This is exactly what procedure P1 does and, hence, procedure P1 is the correct one. Note that under this procedure the partial derivatives of the long run equilibrium stock of the durable good are obtained from (2.46) as

$$\frac{\partial c^*}{\partial h^*} = \frac{\partial [y^* \Gamma(\phi^*)]}{\partial h^*} \qquad \text{for} \qquad h \in \{y, \phi\} . \qquad (2.58)$$

In contrast, procedure P2 does not use the correct long run market equilibrium conditions. It totally differentiates (2.54)–(2.56) and, in doing so, treats ν_1 and ν_2 as exogenously given. According to Proposition 2.4, this procedure ignores that for (2.54)–(2.56) to be the correct long run market equilibrium conditions, we have to set ν_1 and ν_2 equal to long run equilibrium output y^* and product durability ϕ^*. In the comparative static analysis, ν_1 and ν_2 are therefore endogenous with the consequence that procedure P2 is not exact.

One argument in defense of procedure P2 might be to understand its results as approximation for the exact results obtained from procedure P1. However, we then have to check the fitness of this approximation. Instead of the exact derivatives in (2.58), procedure P2 uses

$$\frac{\partial c^*}{\partial h^*} = \frac{\partial [y^* \Psi(\phi^*)]}{\partial h^*} \qquad \text{for} \qquad h \in \{y, \phi\} .$$

Hence, the approximation error made by procedure P2 may be measured by the term

$$\Delta := \Gamma(\phi^*) - \Psi(\phi^*) = \int_0^\infty D(a, \phi^*)(1 - e^{-\delta a}) da .$$

According to this equation, the approximation error is the smaller, the smaller is the discount rate δ. But the error never vanishes since we suppose δ to be strictly positive. Carrying out a comparative static analysis with respect to the number of firms, n, will assess the significance of the approximation error. For this, denote the comparative static results obtained from procedure Pk by the subscript Pk, $k = 1, 2$, and let $|J^*|$ be the Jacobian determinant of the employed long run equilibrium conditions. We then obtain

Proposition 2.5. *Suppose $|J^*| \neq 0$ under both procedures. Then*

$$\left(\frac{\partial \phi^*}{\partial n}\right)_{P1} = \left(\frac{\partial \phi^*}{\partial n}\right)_{P2} = 0 \, , \tag{2.59}$$

whereas it can not be excluded that

$$\left(\frac{\partial y^*}{\partial n}\right)_{P1} \neq \left(\frac{\partial y^*}{\partial n}\right)_{P2} \qquad and \qquad \left(\frac{\partial c^*}{\partial n}\right)_{P1} \neq \left(\frac{\partial c^*}{\partial n}\right)_{P2} \, , \tag{2.60}$$

in general.

Proof. Totally differentiating (2.46)–(2.48) in case of procedure P1 and (2.54)–(2.56) in case of procedure P2 yields

$$\underbrace{\begin{pmatrix} K_{yy} & K_{y\phi} - \dfrac{K_\phi}{y^*} & -\mathcal{X}\Psi \\[2ex] K_{y\phi} - \dfrac{K_\phi}{y^*} & K_{\phi\phi} - y^*(P + c^*P')\Psi'' & -y^*\mathcal{X}\Psi' \\[2ex] C_y & C_\phi & 1 \end{pmatrix}}_{=: \, J^*} \begin{pmatrix} dy^* \\[2ex] d\phi^* \\[2ex] dc^* \end{pmatrix}$$

$$= \begin{pmatrix} c^*(P' + c^*P'')\Psi \\[2ex] y^*c^*(P' + c^*P'')\Psi' \\[2ex] 0 \end{pmatrix} dn \, ,$$

where $\mathcal{X} := [(n+1)P' + nc^*P''] < 0$. In case of procedure P1, we have $C_y = -\Gamma$ and $C_\phi = -y^*\Gamma'$ whereas procedure P2 implies $C_y = -\Psi$ and $C_\phi = -y^*\Psi'$. The Jacobian determinant reads

$$|J^*| := \left[K_{yy} - \mathcal{X}\Psi C_y\right]\left[K_{\phi\phi} - y^*(P + cP')\Psi'' - y^*\mathcal{X}\Psi'C_\phi\right]$$

$$- \left[K_{y\phi} - \frac{K_\phi}{y^*} - y^*\mathcal{X}\Psi'C_y\right]\left[K_{y\phi} - \frac{K_\phi}{y^*} - \mathcal{X}\Psi C_\phi\right] \, .$$

The sign of $|J^*|$ is indeterminate, in general. Besides other factors it depends on C_y and C_ϕ and therefore on the procedure underlying the comparative static analysis. Applying Cramer's rule to the above matrix equation, taking into account $|J^*| \neq 0$ (by assumption) and rearranging terms gives

$$\frac{\partial \phi^*}{\partial n} = \frac{c^* K_y}{|J^*|} \frac{P' + c^* P''}{P + c^* P'} \left[K_{yy} \frac{K_\phi}{K_y} - \left(K_{y\phi} - \frac{K_\phi}{y^*} \right) \right], \tag{2.61}$$

$$\frac{\partial y^*}{\partial n} = \frac{c^* (P' + c^* P'')}{|J^*|} \left[\left[K_{\phi\phi} - y^* (P + c^* P') \Psi'' \right] \right.$$

$$\left. - y^* \left(K_{y\phi} - \frac{K_\phi}{y^*} \right) \Psi' \right], \tag{2.62}$$

$$\frac{\partial c^*}{\partial n} = \frac{c^* (P' + c^* P'')}{|J^*|} \left[\left[K_{\phi\phi} - y^* (P + c^* P') \Psi'' \right] \Psi C_y + y^* K_{yy} \Psi' C_\phi \right.$$

$$\left. - \left(K_{y\phi} - \frac{K_\phi}{y^*} \right) \left(y^* \Psi' C_y + \Psi C_\phi \right) \right]. \tag{2.63}$$

The Abel condition (2.17) implies $K_{yy} K_\phi / K_y = K_{y\phi} - K_\phi / y^*$. Taking this into account in (2.61) proves (2.59). In contrast, the impact of n on y^* and c^* depends on $|J^*|$, C_y and C_ϕ. Consequently, under procedure P1 the derivatives dy^*/dn and dc^*/dn in (2.62) and (2.63) generally take on values different from those under procedure P2. This completes the proof of Proposition 2.5.   □

(2.59) in Proposition 2.5 represents the Swan independence result which we already discussed in the introduction. It says that long run product durability is independent of the number of firms operating in the industry and thus independent of the market structure. The intuition behind this result becomes evident if we write the long run equilibrium condition (2.47) and (2.48) as

$$\frac{\eta_{Ky}}{\eta_{K\phi}} = \frac{1}{\eta_{\Psi\phi}},$$

where $\eta_{\Psi\phi} := \phi \Psi'(\phi) / \Psi(\phi)$ defines the elasticity of the discounted use potential with respect to durability. This equation is independent of demand conditions in the durable good market and, thus, firms choose product durability such that it minimizes production costs. Moreover, both sides of the above conditions do not depend on the firms' output according to the Abel condition (2.17). The cost-minimizing product durability is therefore independent of the firms' market share and independent of market structure.

Swan's independence result was first proved by Sieper and Swan (1973) and then generalized by Abel (1983) who assumed a more general production cost function satisfying (2.17).[14] The new insight of Proposition 2.5 is that Swan's

---

[14] Sieper and Swan (1973) and Abel (1983) prove the Swan result by comparing the two polar market structures of monopoly and perfect competition. The comparative statics reported in our Proposition 2.5 are more general than those of the

result is valid for both procedures P1 and P2. Consequently, with respect to long run product durability, procedure P2 provides a good approximation. Things are different with respect to long run equilibrium output and stock of the durable good. According to (2.60), the impact of the number of firms on these equilibrium values under procedure P2 is not the same as under procedure P1, in general. Thus, the results of procedure P2 do not represent a good approximation for the pertinent results of the exact procedure P1.

In order to avoid incorrect results, the subsequent analysis employs procedure P1. An exception will be Sect. 3.1 in which we reproduce the analysis of Goering and Boyce (1999) who apply procedure P2. However, it will also be examined if and how their results change when the correct procedure P1 is used.

---

authors mentioned above since they include the case of durable goods oligopoly. The oligopoly case is also considered by Goering (1992) who, however, explores a two-period oligopoly model with a cost function linear in output.

3

Regulating Production Emissions of Consumer Durables

This chapter extends the basic model of Sect. 2.2 in order to analyze environmental policy for consumer durables. We will focus on emissions generated during the process of producing the durable goods. Such emissions are important in many countries. For example, the study of Statistikbundesamt (2001) shows that in 1998, the German manufacturing industries generate about 44055 Mio. m^3 of wastewater and about 672 Mio. tons of air emissions like CO_2, NO_x or SO_2. The main part of these emissions arises in sectors producing durable goods or intermediate goods like e.g. metals or plastics which are employed in the production of durable goods.

Goering and Boyce (1999) (hereafter referred to as G&B) have already presented a durable good model accounting for such production emissions. We therefore report on G&B's model in Sect. 3.1, especially in Sects. 3.1.1 and 3.1.3. Reexamining G&B's analysis will turn out to reveal some technical problems with the existence or specification of the market equilibrium. These problems are specified in Sects. 3.1.2 and 3.1.4. They render some of G&B's results obsolete and consequently leave unanswered some important questions about the appropriate environmental regulation of production emissions in durable goods markets. To fill this gap, Sect. 3.2 modifies G&B's model in order to avoid the technical problems and to offer a richer and more comprehensive policy analysis. In doing so, we apply the procedure already introduced in the introduction: Section 3.2.1 first derives the efficient allocation in the modified model. Section 3.2.2 then proceeds by characterizing the market equilibrium and identifying market failure under laissez faire. In Sect. 3.2.3, regulatory schemes capable of restoring efficiency are derived, and Sect. 3.2.4 considers second-best emissions and output taxation in the polluting durable good industry. In Sect. 3.3, we will assess the results of the theoretical analysis of Sect. 3.2 under the perspective of practical policy. This applied discussion addresses the question whether the tax instruments employed in the theoretical models are feasible from a practical point of view.

3.1 The Model of Goering and Boyce (1999)

3.1.1 Model Description and Market Equilibrium

G&B were the first who investigated environmental aspects of durable consumption goods. Applying the notation of Chap. 2, their model is given by

$$k_i(t) = \kappa[\phi_i(t)]y_i(t) \qquad \forall\, i \in N \ , \tag{3.1}$$

$$c_i(t) = \int_{-\infty}^{t} D[t-v, \phi_i(v)]y_i(v)dv \qquad \forall\, i \in N \ , \tag{3.2}$$

$$p(t) = P[C(t)] \qquad \text{with} \quad C(t) = \sum_{j \in N} c_j(t) \ , \tag{3.3}$$

$$\varepsilon_i(t) = \mathcal{E}[y_i(t), \phi_i(t)] \qquad \forall\, i \in N \ , \tag{3.4}$$

$$e(t) = E[W(t)] \qquad \text{with} \quad W(t) = \sum_{j \in N} \varepsilon_j(t) \ . \tag{3.5}$$

Observe that the model consists again of n identical firms. At time t, firm i manufactures $y_i(t)$ units of the durable good with built-in product durability $\phi_i(t)$. Production costs are specified in (3.1). G&B assume the cost function is linear in output. Unit production costs are therefore a function $\kappa : \mathbb{R}_+ \mapsto \mathbb{R}_+$ of product durability only. κ is supposed to be twice continuously differentiable and increasing in product durability, i.e. $\kappa'(\phi) > 0$ for all $\phi \in \mathbb{R}_+$. For the time being, G&B leave the sign of the second derivative of κ unspecified and discuss different cases later. However, note that the assumptions about production costs are consistent with our assumption A3 introduced on p. 24 only if $\kappa''(\phi) \geq 0$ for all $\phi \in \mathbb{R}_+$.

Equation (3.2) represents firm i's stock of the durable good at time t which consists of the remaining units of all vintages produced prior to time t. With respect to the decay function D, G&B suppose assumption A1 from p. 15. Hence, the discounted and undiscounted use potential of the durable good, $\Gamma(\phi)$ and $\Psi(\phi)$, satisfy the conditions derived in Lemma 2.1i on p. 16. $p(t)$ in equation (3.3) denotes the price of the durable's services at time t as a function of the industry stock $C(t)$ at time t. With respect to the demand function, G&B impose assumption A4 introduced on p. 25.

So far the model (3.1)–(3.5) coincides with the basic oligopoly model presented in Sect. 2.2. The difference to the basic model is that model (3.1)–(3.5) takes into account production emissions. The emissions are modeled by the twice continuously differentiable emissions function $\mathcal{E} : \mathbb{R}_+^2 \mapsto \mathbb{R}_+$ in (3.4). According to this function, firm i's production at time t generates emissions $\varepsilon_i(t)$ depending on both output and product durability at time t. As already argued in the introduction on p. 6, an increase in output and/or durability usually requires more material input in the production process. This results in, for example, more smoke, wastewater or solid production waste and so raises production emissions. The emissions function is therefore supposed to be non-decreasing in output and durability, i.e. $\mathcal{E}_y(y, \phi) \geq 0$ and $\mathcal{E}_\phi(y, \phi) \geq 0$ for all $(y, \phi) \in \mathbb{R}_+^2$. G&B include the equality sign since this allows investigating cases in which emissions depend exclusively on output or product durability, respectively.

$W(t)$ in (3.5) measures total industry emissions at time t. Pollutants are assumed to be homogenous so that $W(t)$ equals the sum of emissions from all firms $j \in N$. Industry emissions cause environmental damage. For example, industry smoke contributes to climate change, wastewater pollutes the ground water and solid production waste has to be landfilled and pollutes the environment through e.g. leakage. All these damages are reflected by the damage function $E : \mathbb{R}_+ \mapsto \mathbb{R}_+$ in (3.5). E is twice continuously differentiable and satisfies $E'(W) > 0$ and $E''(W) \geq 0$ for all $W \in \mathbb{R}_+$. Hence, an increase in industry emissions raises the environmental damage at non-decreasing rates.

Having completed the model description, we now investigate a durable good rental game. The game is almost the same as the one presented in Sect. 2.2.1 on pp. 25. The only difference is that G&B suppose the government imposes an emissions tax which firms have to pay for every unit of production emissions. The emissions tax rate τ_ε is assumed to be independent of time. Firm i's instantaneous rental profits then equal rental revenues $p(t)c_i(t)$ less production $k_i(t)$ costs and emissions tax payments $\tau_\varepsilon \varepsilon_i(t)$. The pay-off of firm i is a function $\Pi^{ri} : \mathcal{S}^n \mapsto \mathbb{R}$ where

$$\Pi^{ri}(s_i, s_{-i}) = \int_0^\infty \left[P\left[\sum_{j \in N} \int_{-\infty}^t D[t - v, \phi_j(v)] y_j(v) dv \right] \times \right.$$

$$\times \int_{-\infty}^t D[t - v, \phi_i(v)] y_i(v) dv - \kappa[\phi_i(t)] y_i(t)$$

$$\left. - \tau_\varepsilon \mathcal{E}[y_i(t), \phi_i(t)] \right] e^{-\delta t} dt \qquad (3.6)$$

is the present value of firm i's rental profits over the planning period $[0, \infty[$. As in the basic model presented in Sect. 2.2, G&B focus on the case of open-loop information pattern and derive the open-loop rental equilibrium as defined in definition D2 on p. 27. Similar to Proposition 2.2 on p. 33, it is straightforward to show that the open-loop sales equilibrium is identical to the open-loop rental equilibrium if assumption A5 or assumption A6 is satisfied.

Determining the open-loop rental equilibrium requires solving firm i's profit maximization problem. Firm i chooses the time paths of output and product durability in order to maximize the present value of rental profits in (3.6). In doing so, it takes as given the time paths chosen by its competitors. With the help of (3.1)–(3.4), this maximization problem can be written as

$$\max_{\substack{y_i(0,\infty) \\ \phi_i(0,\infty)}} \int_0^\infty \left[c_i(t) P[c_i(t) + C^i(t)] - \kappa[\phi_i(t)] y_i(t) \right.$$

$$\left. - \tau_\varepsilon \mathcal{E}[y_i(t), \phi_i(t)] \right] e^{-\delta t} dt \right\} \quad (3.7)$$

$$\text{subject to} \quad c_i(t) = \int_{-\infty}^t D[t - v, \phi_i(v)] y_i(v) dv \;,$$

where $C^i(t) = \sum_{j \in N, j \neq i} c_j(t)$ represents the durable stock of firm i's competitors at time t. (3.7) is an optimal control problem with an integral state constraint. The control variables are $y_i(t)$ and $\phi_i(t)$, and $c_i(t)$ represents the state variable. In order to characterize the solution to this problem, we introduce the present-value Hamiltonian

$$\mathcal{H}[y_i(t), \phi_i(t), c_i(t), \lambda_i(t, \infty), t] = \left[c_i(t) P[C(t)] - \kappa[\phi_i(t)] y_i(t) \right.$$

$$\left. - \tau_\varepsilon \mathcal{E}[y_i(t), \phi_i(t)] \right] e^{-\delta t} + \int_t^\infty D[v - t, \phi_i(t)] y_i(t) \lambda_i(v) dv \;, \quad (3.8)$$

where $\lambda_i(t)$ is the costate variable or the shadow price of firm i's stock of the durable good at time t. The Hamiltonian measures the impact of time t control and state variables on the objective function in problem (3.7). As in our basic durable good model presented in Sect. 2.2, changes in $c_i(t)$ alter revenues at time t and changes in $y_i(t)$ and $\phi_i(t)$ influence production costs at t and the stock of the durable good at times $v \geq t$. In contrast to our basic oligopoly model, however, time t control variables now exert an additional effect on rental profits through determining emissions tax payments at time t. This impact is reflected by the term $\tau_\varepsilon \mathcal{E}[y_i(t), \phi_i(t)]$ in the Hamiltonian (3.8).

Applying the maximum principle of Bakke (1974) to firm i's profit maximization (3.7) immediately yields

Proposition 3.1. *Let $[y_i^*(0, \infty), \phi_i^*(0, \infty)]$ be the solution to (3.7) and let $c_i^*(0, \infty)$ be the associated path of the state variable satisfying the integral state constraint in (3.7). Then there exists a path $\lambda_i(0, \infty)$ such that*

$$[y_i^*(t), \phi_i^*(t)] = \arg \max_{y_i(t), \phi_i(t)} \mathcal{H}[y_i(t), \phi_i(t), c_i^*(t), \lambda_i(t, \infty), t] \;, \quad (3.9)$$

$$\lambda_i(t) = \mathcal{H}_c[y_i^*(t), \phi_i^*(t), c_i^*(t), \lambda_i(t, \infty), t]$$

$$= \left[P[C(t)] + c_i^*(t) P'[C(t)] \right] e^{-\delta t} \;, \quad (3.10)$$

for all $t \in [0, \infty[$.

The interpretation of conditions (3.9) and (3.10) is similar to the interpretation of the corresponding conditions in the basic model of Sect. 2.2: In order to attain maximum profits, firm i has to set output and product durability such that they maximize the impact on rental profits and the stock of durable good such that the shadow price of this stock equals marginal revenues.

Observe that the maximizers $[y_i^*(t), \phi_i^*(t)]$ defined in (3.9) may be located at the boundary or in the interior of the domain of the Hamiltonian. Our subsequent analysis will show that each of these cases is relevant under certain conditions. Yet G&B restrict their analysis to constellations where $[y_i^*(t), \phi_i^*(t)]$ is an interior solution for all $t \in [0, \infty[$. Moreover, when they draw conclusions on that basis, they fail to make sure that presupposing an interior solution was warranted in the first place. To show this, our next step is to follow G&B's line of analysis. Hence, we assume for the time being that $[y_i^*(t), \phi_i^*(t)]$ from (3.9) are interior maximizers of the Hamiltonian satisfying the FOCs

$$\mathcal{H}_y(t) = -\Big[\kappa(t) + \tau_\varepsilon \mathcal{E}_y(t)\Big]e^{-\delta t} + \int_t^\infty D(v)\lambda_i(v)dv = 0 \ , \tag{3.11}$$

$$\mathcal{H}_\phi(t) = -\Big[y_i^*(t)\kappa'(t) + \tau_\varepsilon \mathcal{E}_\phi(t)\Big]e^{-\delta t} + y_i^*(t) \int_t^\infty D_\phi(v)\lambda_i(v)dv = 0 \ , \tag{3.12}$$

for all $t \in [0, \infty[$. Inserting (3.10) into these conditions yields

$$\kappa(t) + \tau_\varepsilon \mathcal{E}_y(t) = \int_t^\infty D(v)[P(v) + c_i^*(v)P'(v)]e^{-\delta(v-t)}dv \ , \tag{3.13}$$

$$\kappa'(t) + \frac{\tau_\varepsilon \mathcal{E}_\phi(t)}{y_i^*(t)} = \int_t^\infty D_\phi(v)[P(v) + c_i^*(v)P'(v)]e^{-\delta(v-t)}dv \ , \tag{3.14}$$

for all $t \in [0, \infty[$. (3.13) states that marginal costs (LHS) equal marginal revenues (RHS) of time t output. Marginal revenues accrue at times $v \geq t$ since a marginal increase in time t output enlarges the stock of the durable good at times $v \geq t$.[1] Marginal costs of time t output consist not only of marginal production costs, but comprise also marginal tax payments since a marginal increase in time t output raises production emissions and the tax bill of firm i at time t. In the same way, (3.14) equalizes marginal costs (LHS) and marginal revenues (RHS) of time t product durability. Marginal revenues result from an expansion of the durable stock at times $v \geq t$. Marginal costs equal marginal production costs and emissions tax payments of product durability (per unit of output). The latter effect arises since production emissions are influenced by both output and product durability of the consumption good.

Taking into account Theorem 2 of Kamien and Muller (1976), conditions (3.13), (3.14) and the integral constraint in (3.7) are not only necessary, but also sufficient for firm i's profit maximum if the maximized Hamiltonian $\mathcal{H}^{\max}[c_i(t), \lambda_i(t, \infty), t] := \max_{y_i(t), \phi_i(t)} \mathcal{H}[y_i(t), \phi_i(t), c_i(t), \lambda_i(t, \infty), t]$

---

[1] Remember the use of the phrase 'at times $v \geq t$' introduced in footnote 5 on p. 28.

is concave in $c_i(t)$ for any given $\lambda_i(t, \infty)$ and all $t \in [0, \infty[$. This concavity condition is proved by the same arguments as in the basic model: The pair $[y_i(t), \phi_i(t)]$ which maximizes the Hamiltonian does not depend on $c_i(t)$ since it is determined by (3.11) and (3.12). We therefore obtain $\mathcal{H}_{cc}^{\max}(t) = \mathcal{H}_{cc}(t) = [2P'(t) + c_i(t)P''(t)]e^{-\delta t} < 0$ for all $t \in [0, \infty[$ with the consequence that (3.13), (3.14) and the integral constraint in (3.7) are necessary and sufficient for the solution of firm i's profit maximization problem.

Following definition D2, a market equilibrium is reached if all firms simultaneously attain a profit maximum and no firm has an incentive to deviate from its chosen strategy. G&B focus on a symmetric equilibrium. Furthermore, they consider the stationary state. If we suppose that assumption A9 from p. 38 is satisfied and if we employ definition D5, then the conditions for an (asymptotic) stationary state of the market equilibrium are derived from (3.13), (3.14) and the integral constraint in (3.7) as[2]

$$\kappa(\phi^*) + \tau_\varepsilon \mathcal{E}_y(y^*, \phi^*) = [P(nc^*) + c^* P'(nc^*)] \Psi(\phi^*) , \qquad (3.15)$$

$$\kappa'(\phi^*) + \frac{\tau_\varepsilon \mathcal{E}_\phi(y^*, \phi^*)}{y^*} = [P(nc^*) + c^* P'(nc^*)] \Psi'(\phi^*) , \qquad (3.16)$$

$$c^* = y^* \Gamma(\phi^*) , \qquad (3.17)$$

where y^*, ϕ^* and c^* denote the long run equilibrium values of output, product durability and stock of the durable good, respectively.

According to (3.15)–(3.17), the equilibrium values are functions of the tax rate τ_ε and the number of firms n. In carrying out a comparative static analysis of the long run equilibrium, G&B apply procedure P2 described on p. 43. Although we demonstrated in Sect. 2.2.3 that this procedure does not provide exact results, in general, we follow G&B's analysis for the time being. In Sect. 3.1.4, we will investigate how the results change if we use the exact procedure P1. With the help of procedure P2, comparative static results are obtained by totally differentiating (3.15), (3.16) and the constraint $c^* = y^* \Psi(\phi^*) - \nu_1 \Psi(\nu_2) + \nu_1 \Gamma(\nu_2)$ where ν_1 and ν_2 are taken as given. The relevant comparative static results for the subsequent analysis are

$$\frac{\partial \phi^*}{\partial n} = \frac{c^*(P' + c^* P'')}{|J^*|} \left[\tau_\varepsilon y^* \Psi' \mathcal{E}_{yy} + \frac{\tau_\varepsilon \Psi \mathcal{E}_\phi}{y^*} \left(1 - \eta_{\mathcal{E}_\phi y} \right) \right] , \qquad (3.18)$$

$$\frac{\partial \phi^*}{\partial \tau_\varepsilon} = \frac{1}{|J^*|} \left[\mathcal{X}[\Psi^2 \mathcal{E}_\phi - y^* \Psi \Psi' \mathcal{E}_y] - \tau_\varepsilon \mathcal{E}_{yy} \mathcal{E}_\phi - \frac{\tau_\varepsilon \mathcal{E}_y \mathcal{E}_\phi}{y^*} \left(1 - \eta_{\mathcal{E}_\phi y} \right) \right] , \quad (3.19)$$

---

[2] G&B use definition D3 instead of definition D5. As already explained in Sect. 2.2.2, however, the long run equilibrium conditions are the same for both definitions.

$$\frac{\partial y^*}{\partial \tau_\varepsilon} = \frac{1}{|J^*|} \left[\mathcal{X}[y^{*2}\Psi'^2 \mathcal{E}_y - y^* \Psi \Psi' \mathcal{E}_\phi] - \mathcal{E}_y[y^* \kappa'' + \tau_\varepsilon \mathcal{E}_{\phi\phi} \right.$$

$$\left. - y^* \Psi''(P + c^*P')] - \frac{\tau_\varepsilon \mathcal{E}_\phi^2}{y^*}\left(1 - \eta_{\mathcal{E}_\phi y}\right) \right], \qquad (3.20)$$

where $\mathcal{X} := [(n+1)P' + nc^*P''] < 0$ due to assumption A4 and where $\eta_{\mathcal{E}_\phi y} := y\mathcal{E}_{\phi y}/\mathcal{E}_\phi$ is the elasticity of marginal durability emissions with respect to output. The Jacobian determinant is

$$|J^*| = \tau_\varepsilon \mathcal{E}_{yy}\left[y^* \kappa'' + \tau_\varepsilon \mathcal{E}_{\phi\phi} - y^* \Psi''(P + c^*P')\right] - \frac{\tau_\varepsilon^2 \mathcal{E}_\phi^2}{y^{*2}}\left[1 - \eta_{\mathcal{E}_\phi y}\right]^2$$

$$- \tau_\varepsilon y^{*2} \Psi'^2 \mathcal{X} \mathcal{E}_{yy} - \Psi^2 \mathcal{X}\left[y^* \kappa'' + \tau_\varepsilon \mathcal{E}_{\phi\phi} - y^* \Psi''(P + c^*P')\right]$$

$$- 2\tau_\varepsilon \Psi \Psi' \mathcal{X} \mathcal{E}_\phi\left[1 - \eta_{\mathcal{E}_\phi y}\right]. \qquad (3.21)$$

Without further specification of the model, the sign of $|J^*|$ is indeterminate, in general. However, $|J^*|$ can be shown to be positive in all relevant cases of the subsequent analysis.

To present the conclusions G&B draw from (3.18)–(3.20), we introduce the following conditions.

Condition C1. *The market equilibrium is such that $[y_i^*(0,\infty), \phi_i^*(0,\infty)]$ satisfy (3.9) as interior maximizers determined by (3.11) and (3.12) for all $i \in N$.*

Condition C2. $\tau_\varepsilon = 0$.

Condition C3. $\tau_\varepsilon > 0$ and $\mathcal{E}_\phi(y, \phi) = 0$, $\mathcal{E}_{yy}(y, \phi) < 0$ for all $(y, \phi) \in \mathbb{R}_+^2$.

Condition C4. $\tau_\varepsilon > 0$ and $\mathcal{E}_\phi(y, \phi) = 0$, $\mathcal{E}_{yy}(y, \phi) > 0$ for all $(y, \phi) \in \mathbb{R}_+^2$.

Condition C5. $\tau_\varepsilon > 0$ and $\mathcal{E}_y(y, \phi) = 0$ for all $(y, \phi) \in \mathbb{R}_+^2$.

Condition C6. $\tau_\varepsilon > 0$ and $\mathcal{E}_\phi(y, \phi) = 0$ for all $(y, \phi) \in \mathbb{R}_+^2$.

Condition C1 was presupposed on the previous pages in order to obtain the comparative static results (3.18)–(3.20). By taking into account conditions C2–C6 in (3.18) and (3.19), G&B immediately obtain the following two Propositions.

Proposition 3.2. (Goering and Boyce, 1999, Proposition 1) *In the long run market equilibrium*

(i) $\dfrac{\partial \phi^*}{\partial n} = 0$ *if*   *conditions C1 and C2 hold,*

(ii) $\dfrac{\partial \phi^*}{\partial n} > 0$ *if*   *conditions C1 and C3 hold,*

(iii) $\dfrac{\partial \phi^*}{\partial n} < 0$ *if*   (a) *conditions C1 and C4 hold*

                       or   (b) *conditions C1 and C5 hold.*

Proposition 3.3. (Goering and Boyce, 1999, Proposition 2) *In the long run market equilibrium*

(i) $\dfrac{\partial \phi^*}{\partial \tau_\varepsilon} < 0$ *if*   *conditions C1 and C5 hold,*

(ii) $\dfrac{\partial \phi^*}{\partial \tau_\varepsilon} > 0$ *if*   *conditions C1 and C6 hold.*

As already mentioned above, G&B do not prove whether presupposing an interior solution is warranted in the first place, i.e. they do not check whether conditions C2–C6 are compatible with condition C1. In fact, we will now show that some of the conditions C2–C6 are incompatible with condition C1 so that the associated cases in Proposition 3.2 and 3.3 are empty.

3.1.2 Market Equilibrium Revised

The reason for the incompatibility will be that, under some of the conditions C2–C6, the Hamiltonian (3.8) is either strictly convex or linear in one of the control variables with the consequence that condition C1 is not satisfied, i.e. $[y_i^*(0,\infty), \phi_i^*(0,\infty)]$ from (3.9) either does not exist at all or satisfies (3.9) not as interior maximizer determined by (3.11) and (3.12). To be more explicit, the curvature of the Hamiltonian is determined by the second partial derivatives

$$\mathcal{H}_{yy}(t) = -\tau_\varepsilon \mathcal{E}_{yy}(t) e^{-\delta t} , \tag{3.22}$$

$$\mathcal{H}_{\phi\phi}(t) = -\left[y_i(t)\kappa''(t) + \tau_\varepsilon \mathcal{E}_{\phi\phi}(t) \right] e^{-\delta t} + y_i(t) \int_t^\infty D_{\phi\phi}(v) \lambda_i(v) dv , \tag{3.23}$$

and by

$$\mathcal{H}_{yy}(t)\mathcal{H}_{\phi\phi}(t) - \mathcal{H}_{y\phi}^2(t)$$

$$= \tau_\varepsilon \mathcal{E}_{yy}(t) \left[y_i(t)\kappa''(t) + \tau_\varepsilon \mathcal{E}_{\phi\phi}(t) - y_i(t) \int_t^\infty D_{\phi\phi}(v)\lambda_i(v)e^{\delta t} dv \right] e^{-2\delta t}$$

$$- \left[\kappa'(t) + \tau_\varepsilon \mathcal{E}_{y\phi}(t) - \int_t^\infty D_\phi(v)\lambda_i(v)e^{\delta t} dv \right]^2 e^{-2\delta t} . \tag{3.24}$$

Note that the signs of (3.22)–(3.24) characterize global properties of the Hamiltonian if they hold for all $[y_i(t), \phi_i(t)] \in \mathbb{R}_+^2$. With the help of (3.22)–(3.24), the following proposition can be proved.

Proposition 3.4. *Condition C1 is incompatible with conditions C2, C3 and C5.*

Proof. Let us start with proving the incompatibility of the conditions C1 and C3. Under condition C3, the Hamiltonian is globally and strictly convex in output for all $t \in [0, \infty[$ since $\tau_\varepsilon > 0$ and $\mathcal{E}_{yy}(y, \phi) < 0$ for all $(y, \phi) \in \mathbb{R}_+^2$ imply $\mathcal{H}_{yy}(t) = -\tau_\varepsilon \mathcal{E}_{yy}(t) e^{-\delta t} > 0$ for all $[y_i(t), \phi_i(t)] \in \mathbb{R}_+^2$ and all $t \in [0, \infty[$ according to (3.22). Moreover, the first derivative of the Hamiltonian (3.8) with respect to output becomes

$$\mathcal{H}_y(t) = \underbrace{\int_t^\infty D[v - t, \phi_i(t)] \lambda_i(v) dv - \kappa[\phi_i(t)] e^{-\delta t}}_{=:G[t, \phi_i(t)]} - \underbrace{\tau_\varepsilon \mathcal{E}_y[y_i(t)] e^{-\delta t}}_{<0} . \quad (3.25)$$

Although the properties of the function $G[t, \phi_i(t)]$ are not known, this function can be used to prove the incompatibility of the conditions C1 and C3. We have to distinguish two cases:

1. Suppose there exists a time point $t_1 \in [0, \infty[$ with $G[t_1, \phi_i(t_1)] > 0$ for some $\phi_i(t_1) \geq 0$. If we set $\phi_i(t_1)$ such that $G[t_1, \phi_i(t_1)] > 0$, then $\tau_\varepsilon > 0$ and (3.25) imply

    $$\mathcal{H}_y(t_1) = 0 \quad \Leftrightarrow \quad \mathcal{E}_y[y_i(t_1)] e^{-\delta t_1} = G[t_1, \phi_i(t_1)]/\tau_\varepsilon > 0 .$$

    Taking durability $\phi_i(t_1)$ as given, the latter expression has exactly one solution for $y_i(t_1)$ since $\mathcal{E}_y[y_i(t_1)]$ is monotone. This together with the global and strict convexity of the Hamiltonian in output renders $\mathcal{H}(t_1)$ U-shaped with respect to $y_i(t_1)$. Hence, for given $\phi_i(t_1)$, the Hamiltonian $\mathcal{H}(t_1)$ can always be increased by raising $y_i(t_1)$ and there does not exist a $[y_i(t_1), \phi_i(t_1)]$ which maximizes $\mathcal{H}(t_1)$ as required by (3.9). Because (3.9) is necessary for the solution of (3.7), there does not exist a profit maximum of firm i, i.e. there do not exist optimal time paths $[y_i^*(0, \infty), \phi_i^*(0, \infty)]$ and condition C1 is violated.

2. Suppose for all $t \in [0, \infty[$ we have $G[t, \phi_i(t)] \leq 0$ for all $\phi_i(t) \geq 0$. Then (3.25) implies $\mathcal{H}_y(t) < 0$ for all $y_i(t) \geq 0$ and all $t \in [0, \infty[$. The Hamiltonian is therefore maximized at $y_i(t) = 0$ and firm i's profit-maximizing strategy is to abstain from producing the durable good, i.e. $y_i^*(t) = 0$ and $\phi_i^*(t) = 0$ for all $t \in [0, \infty[$. Consequently, condition C1 is violated since we have no interior solution.

These two cases are exhausting, i.e. there are no other cases which we have to consider. Since in both cases condition C1 is violated, it follows that conditions C1 and C3 are incompatible.[3]

---

[3] This result is plausible. $\mathcal{H}_y(t)$ may be interpreted as the present value of marginal profits a unit of vintage t generates over its whole lifetime. So, in case 1, this

Now turn to proving the incompatibility of conditions C1 and C2. For $\tau_\epsilon = 0$, the Hamiltonian (3.8) simplifies to

$$\mathcal{H}(t) = c_i(t)P[C(t)]e^{-\delta t} + y_i(t)G[t, \phi_i(t)] , \qquad (3.26)$$

where $G[t, \phi_i(t)]$ is defined in (3.25). We now distinguish three exhausting cases.

1. Suppose there exists a $t_2 \in [0, \infty[$ for which the function $G[t_2, \phi_i(t_2)]$ has no maximum with respect to $\phi_i(t_2)$. Then also the Hamiltonian in (3.26) has no maximum for $t = t_2$ and the necessary condition (3.9) for the solution of (3.7) is violated. Consequently, it does not exist a solution $[y_i^*(0, \infty), \phi_i^*(0, \infty)]$ to (3.7) and condition C1 is violated.

2. Suppose that for all $t \in [0, \infty[$, the function $G[t, \phi_i(t)]$ possesses a maximum with respect to $\phi_i(t)$ and this maximum is positive for at least one time point $t_3 \in [0, \infty[$. Setting $\phi_i(t_3)$ such that $G[t_3, \phi_i(t_3)]$ attains its maximum and increasing $y_i(t_3)$ to infinity shows that the Hamiltonian (3.26) does not attain a maximum for $t = t_3$. Hence, there does not exist a solution $[y_i^*(0, \infty), \phi_i^*(0, \infty)]$ of (3.7) and condition C1 is violated.

3. Suppose that for all $t \in [0, \infty[$, the function $G[t, \phi_i(t)]$ possesses a maximum with respect to $\phi_i(t)$ and this maximum is non-positive. If the maximum of $G[t, \phi_i(t)]$ is negative, then the Hamiltonian (3.26) attains a maximum at $y_i^*(t) = \phi_i^*(t) = 0$. In this case, condition C1 is violated since we have no interior solution. If the maximum of $G[t, \phi_i(t)]$ is equal to zero, then the Hamiltonian (3.26) attains a maximum by setting $y_i(t) = 0$ and $\phi_i(t)$ arbitrary or $y_i(t)$ arbitrary and $\phi_i(t)$ such that $G[t, \phi_i(t)] = 0$. Hence, in this case, we obtain multiple solutions, and the solution may be characterized by interior values for output and durability, indeed. But these values are not determined by (3.11) and (3.12) as presupposed in condition C1. Consequently, condition C1 is violated.

Condition C1 is violated in all possible cases with the consequence that C1 and C2 are incompatible. Finally, to prove the incompatibility of C1 and C5, note that the Hamiltonian under condition C5 is again linear in output. Hence, we can apply the same arguments as in case of C3. This observation completes the proof of Proposition 3.4.   □

Due to Proposition 3.4, G&B's results in Proposition 3.2i, ii, iiib and in Proposition 3.3i are empty. The results which remain are Proposition 3.2iiia and Proposition 3.3ii. Proposition 3.2iiia states that product durability decreases

---

present value of marginal profits is continuously increasing in output at $t = t_1$ (since the Hamiltonian is U-shaped in output) and firm i takes advantage of this property in order to increase its profits unboundedly. In case 2, the present value of marginal profits is negative for all t and, hence, producing the durable good is not profitable.

with an increasing number of firms if emissions are taxed and if the emissions function is independent of durability and convex in output. This result of G&B's analysis is interesting for two reasons. First, it shows that Swan's independence result does generally not hold in a durable good industry with emissions taxation: If production emissions are taxed, then firms set product durability such that it minimizes production costs *plus* tax payments instead of production costs only. Since the firm's marginal tax payments depend on the firm's output (due to $\tau_\varepsilon \mathcal{E}_{yy} > 0$) and since the firm's output depends on the number of firms operating in the industry, the cost-minimizing product durability is influenced by the market structure and Swan's result does not hold. Second, Proposition 3.2iiia states that durability is *decreasing* in the number of firms implying that a monopolist generates the greatest product durability. This finding refutes the hypothesis of planned obsolescence which blames the monopolist to manufacture less durable products than producers in perfect competition.

Proposition 3.3ii states that an increase in the emissions tax rate increases product durability if the emissions function does not depend on product durability. The reason is obvious. If the tax rate is increased, then it becomes more expensive to pollute the environment. The producers restrict production emissions which, in case of $\mathcal{E}_\phi = 0$, is equivalent to reducing output; condition C6 implies $\partial y^*/\partial \tau_\varepsilon < 0$ according to (3.20). This decline in output exerts a negative effect on the stock of the durable good and on rental revenues. In order to partially compensate for the declining stocks and revenues, the firms increase product durability of the consumption goods.

It should be noted that, under the conditions of Proposition 3.2iiia and Proposition 3.3ii, the Hamiltonian is strictly concave in output, indeed. But excluding that these results are also empty requires presenting (at least) one example in which condition C1 is compatible with conditions C4 and C6, respectively. This may be done by numerically solving the firm's profit maximization problem (3.7). To the best of my knowledge, however, the literature has not yet provided methods for numerically solving optimal control problems with integral constraints. Hence, it has to be left open whether there are model specifications for which C4 and C6 are compatible with condition C1.

3.1.3 Second-Best Emissions Taxation

Besides analyzing the properties of the market equilibrium, the second focus of G&B is to determine the optimal emissions tax rate, i.e. the emissions tax rate which maximizes social welfare in the polluting durable good industry described by (3.1)–(3.5). The welfare-maximizing tax is a second-best tax since it has to account for two market imperfections, the market power of producers and the environmental externality, and hence it fails to implement the (first-best) Pareto-efficient allocation, in general. As already discussed in the introduction on pp. 9, conventional economic intuition suggests that the second-best tax underinternalizes the marginal environmental damage whereas the optimal taxation literature identifies several reasons why the second-best emissions tax may exceed the marginal environmental damage. The intention of G&B is to provide a further rationale for overinternalization, namely product durability as an explicit choice variable of producers.

G&B address this issue by focusing on the long run of the economy. Long run social welfare V is defined as consumer benefits from consuming the services of the durable good, $S(C) = \int_0^C P(\xi)d\xi$, net of production and environmental costs. In maximizing V, G&B account for the adjustment path to the stationary state. As shown in Sect. 2.2.2, this requires taking into account the state constraint in (2.53). Therefore, the long run welfare maximization problem reads

$$\left. \begin{aligned} \max_{\tau_\varepsilon} V(\tau_\varepsilon) &= S[nc^*(\tau_\varepsilon)] - n\kappa[\phi^*(\tau_\varepsilon)]y^*(\tau_\varepsilon) - E\Big[n\mathcal{E}[y^*(\tau_\varepsilon), \phi^*(\tau_\varepsilon)]\Big] \\ \text{subject to} \quad c^*(\tau_\varepsilon) &= y^*(\tau_\varepsilon)\Psi[\phi^*(\tau_\varepsilon)] - \nu_1\Psi(\nu_2) + \nu_1\Gamma(\nu_2) , \end{aligned} \right\} \quad (3.27)$$

where ν_1 and ν_2 are taken as given. V in (3.27) measures social welfare in a symmetric long run market equilibrium of the durable good industry. y^*, ϕ^* and c^* are the long run equilibrium values of output, durability and stock of the durable good, respectively. Provided condition C1 is satisfied, the equilibrium values are determined by the equilibrium condition (3.15)–(3.17) and, thus, they are functions of the effluent charge τ_ε. Social welfare is therefore also a function $V(\tau_\varepsilon)$ of the emissions tax rate. The FOC $V_\tau = 0$ for a long run welfare maximum may be rearranged to yield

$$\left[\Psi P - \kappa - E'\mathcal{E}_y\right]\frac{\partial y^*}{\partial \tau_\varepsilon} + \left[y^*\Psi'P - y^*\kappa' - E'\mathcal{E}_\phi\right]\frac{\partial \phi^*}{\partial \tau_\varepsilon} = 0 . \qquad (3.28)$$

For notational convenience, the arguments of all functions are suppressed. Exploiting the market equilibrium conditions (3.15)–(3.17) in order to replace $\Psi P - \kappa$ and $y^*\Psi'P - y^*\kappa'$ in (3.28) and solving the resulting expression with respect to τ_ε yields the second-best emissions tax rate

$$\tau_\varepsilon^{sb} = E' + c^*P'\left[\Psi\frac{\partial y^*}{\partial \tau_\varepsilon} + y^*\Psi'\frac{\partial \phi^*}{\partial \tau_\varepsilon}\right] \bigg/ \left[\mathcal{E}_y\frac{\partial y^*}{\partial \tau_\varepsilon} + \mathcal{E}_\phi\frac{\partial \phi^*}{\partial \tau_\varepsilon}\right] . \qquad (3.29)$$

The second-best emissions tax rate equals the marginal environmental damage E' plus an additional term which mainly depends on the impact of the emissions tax on equilibrium output and equilibrium product durability, $\partial y^*/\partial\tau_\varepsilon$ and $\partial\phi^*/\partial\tau_\varepsilon$. This impact is determined by the comparative static analysis which we have already conducted in the analysis of the market equilibrium in Sect. 3.1.1. The relevant derivatives are (3.19) and (3.20).

For the purpose of comparison, G&B first consider second-best emissions taxation in the event that product durability is exogenously given. They obtain

Proposition 3.5. (Goering and Boyce, 1999, Proposition 3) *Suppose condition C1 is satisfied and product durability is exogenously given. Then* $\tau_\varepsilon^{sb} < E'$.

Proof. Under condition C1, the second-best emissions tax rate is determined by (3.29). For given product durability, we obtain $\partial\phi^*/\partial\tau_\varepsilon = 0$ with the consequence that (3.29) simplifies to

$$\tau_\varepsilon^{sb} = E' + c^* P' \frac{\Psi}{\mathcal{E}_y}.$$

The RHS of this expression is smaller than E' since $P' < 0$ and $\Psi, \mathcal{E}_y > 0$. □

Hence, G&B obtain the conventional underinternalization result of second-best emissions taxation if producers are not able to vary product durability. The reason is that output represents the only choice variable of producer. Hence, taxing emissions exactly by the marginal damage internalizes the environmental externality, indeed, but also results in an underprovision of the durable good since producers operate under imperfect competition. Lowering the tax rate below marginal damage is welfare-enhancing since it induces firms to increase output thus raising efficiency.

Taking the result of Proposition 3.5 as a benchmark, G&B turn to the case in which product durability is one of the firms' decision variables. Rather than investigating the case with general costs, demand, decay and emissions functions, they focus on some special cases. In order to present G&B's results, we introduce

Condition C7. $P''(C) = 0$ *for all* $C \in \mathbb{R}_+$, $D_{\phi\phi}(a, \phi) = 0$ *for all* $(a, \phi) \in \mathbb{R}_+^2$, $\mathcal{E}_\phi(y, \phi) = 0$ *for all* $(y, \phi) \in \mathbb{R}_+^2$ *and* $\kappa''(\phi) > 0$ *for all* $\phi \in \mathbb{R}_+$.

Condition C8. $P''(C) = 0$ *for all* $C \in \mathbb{R}_+$, $D_{\phi\phi}(a, \phi) = 0$ *for all* $(a, \phi) \in \mathbb{R}_+^2$, $\mathcal{E}_\phi(y, \phi) = 0$ *for all* $(y, \phi) \in \mathbb{R}_+^2$ *and* $\kappa''(\phi) = 0$ *for all* $\phi \in \mathbb{R}_+$.

Condition C9. $P''(C) = 0$ *for all* $C \in \mathbb{R}_+$, $D_{\phi\phi}(a, \phi) = 0$ *for all* $(a, \phi) \in \mathbb{R}_+^2$, $\mathcal{E}_\phi(y, \phi) = 0$ *for all* $(y, \phi) \in \mathbb{R}_+^2$ *and* $\kappa''(\phi) < 0$ *for all* $\phi \in \mathbb{R}_+$.

These conditions state that the demand and decay functions are linear, the emissions function depends on output only and the unit cost function is convex (C7), linear (C8) or concave (C9). By considering conditions C7–C9 in the comparative static results (3.19)–(3.21) and in the second-best emissions tax (3.29), G&B obtain

Proposition 3.6. (Goering and Boyce, 1999, Proposition 5) *In the long run*

(i) $\tau_\varepsilon^{sb} < E'$ *if*   *conditions C1 and C7 hold,*

(ii) $\tau_\varepsilon^{sb} = E'$ *if*   *conditions C1 and C8 hold,*

(iii) $\tau_\varepsilon^{sb} > E'$ *if*   *conditions C1 and C9 hold.*

Based on Proposition 3.6iii, G&B seem to have found a further case of overinternalization: The second-best emissions tax rate in a polluting durable good industry with endogenous product durability exceeds the marginal environmental damage if all firms possess an interior solution to their profit maximization problem, the demand and the decay functions are linear, emissions depend on output only and unit costs are concave in product durability so that it becomes the easier to increase durability the greater durability already is. G&B explain their finding by a third distortion only inherent in durable good markets, namely '... the misallocation due to producers choosing a durability which does not minimize the social cost of providing a given service level.' (G&B, p. 136) In their view, an increase in the emissions tax has not only the disadvantage of further decreasing output below its efficient level, but also possesses the advantage of moving product durability closer to its efficient level. If the latter effect overcompensates the former, then overinternalization becomes second-best optimal.

3.1.4 Second-Best Emissions Taxation Revised

A closer look at previous results of the durability literature raises doubts whether the above overinternalization result of G&B is true. Under laissez faire in a durable good industry *without* pollution, Goering (1992) shows that product durability is independent of the market structure (Swan's independence result) and, thus, it is socially optimal not only under perfect competition, but also in oligopoly. Hence, under laissez faire in a durable good oligopoly *with* pollution, a possible distortion regarding durability can only rest on the environmental externality and does not represent a distinct distortion. This implies that there are only the conventional distortions present and that the second-best emissions tax falls short of marginal damage: If emissions are taxed with a rate equal to the marginal damage, then the first distortion, namely the environmental externality *together with* a possible misallocation of durability, is fully corrected for whereas the second distortion, namely the market power of firms, tends to restrict output and the stock of the durable below their socially optimal levels. According to the standard second-best argument, society can gain from lowering the emissions tax since this reduction shifts output and the stock of the durable closer to their efficient levels. Thus, it is plausible that the second-best tax fall short of marginal damage even in a polluting durable good industry with endogenous product durability.

We will now confirm this intuitive assessment through showing that conditions C1 and C9 are incompatible with the consequence that Proposition 3.6iii is empty.[4] To be more explicit, consider

Proposition 3.7. *Conditions C1 and C9 are incompatible.*

Proof. Under condition C9, we have $P'' = D_{\phi\phi} = \mathcal{E}_\phi = 0$ and $\kappa'' < 0$. Taking into account this set of assumptions in (3.23) yields $\mathcal{H}_{\phi\phi}(t) =$

---

[4] The following argumentation is based on Runkel (2002).

$-y_i(t)\kappa''(t)e^{-\delta t} > 0$ for all $[y_i(t), \phi_i(t)] \in \mathbb{R}^2_+$. Hence, the Hamiltonian is globally and strictly convex in $\phi_i(t)$. Moreover, for $\mathcal{E}_\phi = 0$, equation (3.12) implies

$$\mathcal{H}_\phi(t) = 0 \qquad \Leftrightarrow \qquad \kappa'[\phi_i(t)] = \int_t^\infty D_\phi[v - t, \phi_i(t)]\lambda_i(v)e^{\delta t}dv > 0 \ .$$

Owing to $D_{\phi\phi} = 0$, the RHS of this equation is independent of $\phi_i(t)$ and, hence, the equation has exactly one solution for $\phi_i(t)$ due to the monotony of κ'. This together with the global convexity of $\mathcal{H}$ with respect to $\phi_i(t)$ implies that, for any given $y_i(t)$, the Hamiltonian is U-shaped with respect to $\phi_i(t)$. Therefore, it does not exist a $[y_i(t), \phi_i(t)]$ which solves (3.9) since $\mathcal{H}$ can always be increased by increasing $\phi_i(t)$. Because (3.9) is a necessary condition for the solution of (3.7), there does not exist a solution $[y_i^*(0, \infty), \phi_i^*(0, \infty)]$ to the firm's profit maximization problem and condition C1 is not satisfied.   □

According to Proposition 3.7, the G&B's overinternalization result in Proposition 3.6iii is empty. As the proof of Proposition 3.7 shows, the reason for this result is that, under condition C9, the Hamiltonian is strictly convex in durability implying that there does not exist a solution to the firm's profit maximization problem. The non-existence of the profit maximum implies, in turn, that there exists neither a long run market equilibrium nor a second-best emissions tax.

Second-Best Emissions Tax under Concave Objective Functions

The major aim of G&B is to compare the second-best emissions tax in a durable good industry with that in a nondurable good industry previously investigated by e.g. Barnett (1980) and Ebert (1992). In contrast to G&B, the latter authors consider the case of general demand, costs and emissions functions and assume a concave objective function of the individual firm as well as an interior solution for all variables. A suitable comparison therefore requires assuming analogous conditions for the durable good industry. The assumption of an interior solution is already contained in condition C1. The concavity of the Hamiltonian is specified in

Condition C10. *The Hamiltonian (3.8) is globally and strictly concave in $[y_i(t), \phi_i(t)]$ for all $t \in [0, \infty[$ and all $i \in N$.*

Since this condition ensures the existence of a pair $[y_i^*(t), \phi_i^*(t)]$ satisfying (3.9), it avoids the convexity problems which lead to Proposition 3.4 and 3.7. We then obtain

Proposition 3.8. *Suppose conditions C1 and C10 are satisfied and the emissions function satisfies $\eta_{\mathcal{E}_\phi y} \leq 1$ for all $(y, \phi) \in \mathbb{R}^2_+$. Then $\tau_\varepsilon^{sb} < E'$.*

Proof. The concavity condition C10 is satisfied if and only if $\mathcal{H}_{yy}(t) < 0$, $\mathcal{H}_{\phi\phi}(t) < 0$ and $\mathcal{H}_{yy}(t)\mathcal{H}_{\phi\phi}(t) - \mathcal{H}_{y\phi}^2(t) > 0$ for all $[y_i(t), \phi_i(t)] \in \mathbb{R}_+^2$. Taking into account (3.22)–(3.24) and focusing on the long run market equilibrium yields

$$\tau_\varepsilon \mathcal{E}_{yy} > 0 , \qquad y^* \kappa'' + \tau_\varepsilon \mathcal{E}_{\phi\phi} - y^* \Psi''(P + c^* P') > 0 , \qquad (3.30)$$

$$\tau_\varepsilon \mathcal{E}_{yy} \left[y^* \kappa'' + \tau_\varepsilon \mathcal{E}_{\phi\phi} - y^* \Psi''(P + c^* P') \right] - \frac{\tau_\varepsilon^2 \mathcal{E}_\phi^2}{y^{*2}} \left[1 - \eta_{\mathcal{E}_\phi y} \right]^2 > 0 . \quad (3.31)$$

Condition C1 ensures that the second-best emissions tax rate is captured by (3.29) and that the partial derivatives contained in this tax rate are represented by (3.19) and (3.20). Hence, the bracketed terms in (3.29) become

$$\Psi \frac{\partial y^*}{\partial \tau_\varepsilon} + y^* \Psi' \frac{\partial \phi^*}{\partial \tau_\varepsilon} = -\frac{1}{|J^*|} \left[\Psi \mathcal{E}_y [y^* \kappa'' + \tau_\varepsilon \mathcal{E}_{\phi\phi} - y^* \Psi''(P + c^* P')] \right.$$

$$\left. + y^* \tau_\varepsilon \Psi' \mathcal{E}_\phi \mathcal{E}_{yy} + \frac{\tau_\varepsilon \mathcal{E}_\phi}{y^*} \left(y^* \Psi' \mathcal{E}_y + \Psi \mathcal{E}_\phi \right) \left(1 - \eta_{\mathcal{E}_\phi y} \right) \right] , \qquad (3.32)$$

$$\mathcal{E}_y \frac{\partial y^*}{\partial \tau_\varepsilon} + \mathcal{E}_\phi \frac{\partial \phi^*}{\partial \tau_\varepsilon} = \frac{1}{|J^*|} \left[\mathcal{X}(y^* \Psi' \mathcal{E}_y - \Psi \mathcal{E}_\phi)^2 - \tau_\varepsilon \mathcal{E}_\phi^2 \mathcal{E}_{yy} \right.$$

$$\left. - \frac{2\tau_\varepsilon \mathcal{E}_\phi^2 \mathcal{E}_y}{y^*} \left(1 - \eta_{\mathcal{E}_\phi y} \right) - \mathcal{E}_y^2 [y^* \kappa'' + \tau_\varepsilon \mathcal{E}_{\phi\phi} - y^* \Psi''(P + c^* P')] \right] , \quad (3.33)$$

where the Jacobian determinant $|J^*|$ is that from (3.21). The long run concavity conditions (3.30) and (3.31) together with $\eta_{\mathcal{E}_\phi y} \le 1$ and $\mathcal{X} < 0$ imply that $|J^*|$ is positive and that (3.32) and (3.33) are negative. From (3.29) follows $\tau_\varepsilon^{sb} < E'$. $\qquad \square$

Proposition 3.8 shows that the conditions for the second-best emissions tax in the durable good industry being smaller than the marginal damage are not very restrictive: First, a concave objective function and an interior solution are typically assumed in all economic models which do not explicitly focus on corner solutions or existence problems. Especially, they are used in the second-best taxation models analyzed by Barnett (1980) and Ebert (1992) which serve as a standard of comparison for the present durable good model.[5] Second,

---

[5] Of course, it cannot be excluded that conditions C1 and C10 are incompatible, too. The global concavity of the Hamiltonian ensures the existence of a pair $[y_i^*(t), \phi_i^*(t)]$ satisfying (3.9), indeed, but this pair may be at the boundary. The compatibility of C1 and C10 may be demonstrated by numerically identifying at least one case with concave Hamiltonian and interior solution. As already

the condition $\eta_{\mathcal{E}_\phi y} \leq 1$ is equivalent to $\mathcal{E}_\phi/y^* - \mathcal{E}_{\phi y} \geq 0$ which is satisfied by all separable emissions functions ($\mathcal{E}_{\phi y} = 0$) and by all those non-separable emissions functions where an increase in output reduces marginal emissions of durability ($\mathcal{E}_{\phi y} < 0$) or increases marginal emissions of durability only slightly ($\mathcal{E}_{\phi y} \leq \mathcal{E}_\phi/y^*$).[6] Furthermore, note that the conditions for $\tau_\varepsilon^{sb} < E'$ listed in Proposition 3.8 are sufficient, but not necessary. Hence, the second-best tax may be smaller than the marginal damage even if $\eta_{\mathcal{E}_\phi y} > 1$.

Of course, in case of $\eta_{\mathcal{E}_\phi y} > 1$, it cannot be ruled out that the second-best emissions tax is greater than marginal damage ($\tau_\varepsilon^{sb} > E'$). However, such a case of overinternalization, if it exists, has counterintuitive implications. To see this, note that the second-best emissions tax (3.29) may be written as

$$\tau_\varepsilon^{sb} = E' + c^* P' \frac{\partial c^*/\partial \tau_\varepsilon}{\partial \varepsilon^*/\partial \tau_\varepsilon} , \qquad (3.34)$$

where $\varepsilon^* := \mathcal{E}(y^*, \phi^*)$ represents the firm's emissions in the long run market equilibrium. $\tau_\varepsilon^{sb} > E'$ implies that either the numerator or the denominator in the second term of (3.34) is *positive*. However, the denominator equals the change of the firm's equilibrium emissions due to a marginal change in the emissions tax. There is a strong plausibility for this change to be *negative*. Moreover, the numerator is also expected to be *negative* since it equals the change in the firm's equilibrium stock of the durable good due to a marginal change in the emissions tax. This change to be positive implies that at least one of the equilibrium values, either that of durability or that of output, *increases* as the emissions tax increases. Such an implication is counterintuitive since both variables are positively correlated with the firm's emissions which in this scenario *decrease*. The plausibility of the negative correlation between the emissions tax rate, on the one hand, and the amount of emissions and the durable stock, on the other hand, is supported by the empirical results of Brännlund and Gren (1999) who find that firms restrict their production and emissions when regulated by environmental policy.

Revising the Procedure of Comparative Statics

Following G&B, we have used procedure P2 in carrying out the comparative static analysis of the long run market equilibrium. From the discussion in

---

mentioned on p. 57, however, methods for numerically solving optimal control problems with integral constraints are not yet available in literature. Note also that Barnett (1980) and Ebert (1992) do not explicitly check the compatibility of their assumptions either.

[6] For example, $\mathcal{E}(y, \phi) = G(y) + (\alpha y + \beta)\phi$ with $\alpha, \beta, G', G'' > 0$ seems to be a reasonable emissions function since a marginal increase in durability causes pollution by additional resources needed to make one unit of the good more durable (α) and by additional resources needed for R&D (β). The production costs of these additional resources are reflected in the cost function $\kappa(\phi)y$. For the above emissions function, we obtain $\mathcal{E}_{\phi y} = \alpha > 0$, but $\mathcal{E}_\phi/y^* - \mathcal{E}_{\phi y} = \beta/y^* > 0$.

Sect. 2.2.3 it is known that procedure P2 is not exact, but only represents an approximation. This observation raises the question whether the second-best emissions tax changes if we employ the exact procedure P1. To answer this question, define the elasticity of emissions with respect to output and durability as $\eta_{\mathcal{E}y} := y\mathcal{E}_y/\mathcal{E}$ and $\eta_{\mathcal{E}\phi} := \phi\mathcal{E}_\phi/\mathcal{E}$, respectively. Moreover, recall that $\eta_{\Psi\phi} := \phi\Psi'/\Psi$ is the elasticity of the discounted use potential with respect to product durability. We then obtain

Proposition 3.9. *Suppose comparative statics are carried out with the help of procedure P1 instead of procedure P2. Furthermore, suppose the conditions C1 and C10 are satisfied and the emissions function satisfies $\eta_{\mathcal{E}_\phi y} \leq 1$ and $\eta_{\mathcal{E}\phi} \leq \eta_{\Psi\phi}\eta_{\mathcal{E}y}$ for all $(y,\phi) \in \mathbb{R}^2_+$. Then $\tau_\varepsilon^{sb} < E'$.*

Proof. (3.29) represents the second-best emissions tax also under procedure P1. However, we obtain different expressions for $\partial y^*/\partial\tau_\varepsilon$ and $\partial\phi^*/\partial\tau_\varepsilon$. According to condition C1, the market equilibrium conditions are still (3.15)–(3.17). When we totally differentiate these conditions, we now have to use the derivatives $\partial c^*/\partial y^* = \Gamma(\phi^*)$ and $\partial c^*/\partial\phi^* = y^*\Gamma'(\phi^*)$ instead of $\partial c^*/\partial y^* = \Psi(\phi^*)$ and $\partial c^*/\partial\phi^* = y^*\Psi'(\phi^*)$. The comparative static analysis then yields

$$\frac{\partial\phi^*}{\partial\tau_\varepsilon} = \frac{1}{|J^*|}\left[\frac{\Gamma\Psi\mathcal{X}\mathcal{E}}{\phi^*}\left(\eta_{\mathcal{E}\phi} - \eta_{\Psi\phi}\eta_{\mathcal{E}y}\right) - \tau_\varepsilon\mathcal{E}_{yy}\mathcal{E}_\phi\right.$$

$$\left. - \frac{\tau_\varepsilon\mathcal{E}_y\mathcal{E}_\phi}{y^*}\left(1 - \eta_{\mathcal{E}_\phi y}\right)\right], \quad (3.35)$$

$$\frac{\partial y^*}{\partial\tau_\varepsilon} = \frac{1}{|J^*|}\left[\frac{y^*\Gamma'\Psi\mathcal{X}\mathcal{E}}{\phi^*}\left(\eta_{\Psi\phi}\eta_{\mathcal{E}y} - \eta_{\mathcal{E}\phi}\right) - \mathcal{E}_y[y^*\kappa'' + \tau_\varepsilon\mathcal{E}_{\phi\phi}\right.$$

$$\left. - y^*\Psi''(P + c^*P')] - \frac{\tau_\varepsilon\mathcal{E}_\phi^2}{y^*}\left(1 - \eta_{\mathcal{E}_\phi y}\right)\right], \quad (3.36)$$

where the Jacobian determinant reads

$$|J^*| = \tau_\varepsilon\mathcal{E}_{yy}\left[y^*\kappa'' + \tau_\varepsilon\mathcal{E}_{\phi\phi} - y^*\Psi''(P + c^*P')\right] - \frac{\tau_\varepsilon^2\mathcal{E}_\phi^2}{y^{*2}}\left[1 - \eta_{\mathcal{E}_\phi y}\right]^2$$

$$- \tau_\varepsilon y^{*2}\Gamma'\Psi'\mathcal{X}\mathcal{E}_{yy} - \Gamma\Psi\mathcal{X}\left[y^*\kappa'' + \tau_\varepsilon\mathcal{E}_{\phi\phi} - y^*\Psi''(P + c^*P')\right]$$

$$- \tau_\varepsilon\mathcal{X}\mathcal{E}_\phi\left[\Gamma'\Psi + \Gamma\Psi'\right]\left[1 - \eta_{\mathcal{E}_\phi y}\right].$$

Since $\mathcal{H}$ in (3.8) is assumed to be strictly concave, we obtain the long run concavity conditions (3.30) and (3.31). These conditions jointly with $\mathcal{X} < 0$ and $\eta_{\mathcal{E}_\phi y} \leq 1$ imply $|J^*| > 0$. Taking into account the comparative static results (3.35) and (3.36), the bracketed terms in the second-best emissions tax (3.29) become, respectively,

$$\Psi \frac{\partial y^*}{\partial \tau_\varepsilon} + y^* \Psi' \frac{\partial \phi^*}{\partial \tau_\varepsilon} = -\frac{1}{|J^*|} \left[\Psi \mathcal{E}_y [y^* \kappa'' + \tau_\varepsilon \mathcal{E}_{\phi\phi} - y^* \Psi''(P + c^* P')] \right.$$

$$+ y^* \tau_\varepsilon \Psi' \mathcal{E}_\phi \mathcal{E}_{yy} + \frac{y^* \Psi \mathcal{X} \mathcal{E}}{\phi^*} \left(\Gamma' \Psi - \Gamma \Psi' \right) \left(\eta_{\mathcal{E}_\phi} - \eta_{\Psi_\phi} \eta_{\mathcal{E}_y} \right)$$

$$+ \left. \frac{\tau_\varepsilon \mathcal{E}_\phi}{y^*} \left(y^* \Psi' \mathcal{E}_y + \Psi \mathcal{E}_\phi \right) \left(1 - \eta_{\mathcal{E}_\phi y} \right) \right] < 0 , \qquad (3.37)$$

$$\mathcal{E}_y \frac{\partial y^*}{\partial \tau_\varepsilon} + \mathcal{E}_\phi \frac{\partial \phi^*}{\partial \tau_\varepsilon} = \frac{1}{|J^*|} \left[\frac{\Gamma \Psi \mathcal{X} \mathcal{E}^2}{\phi^{*2}} \left(\eta_{\mathcal{E}_y} \eta_{\Gamma_\phi} - \eta_{\mathcal{E}_\phi} \right) \left(\eta_{\mathcal{E}_y} \eta_{\Psi_\phi} - \eta_{\mathcal{E}_\phi} \right) \right.$$

$$- \tau_\varepsilon \mathcal{E}_\phi^2 \mathcal{E}_{yy} - \mathcal{E}_y^2 [y^* \kappa'' + \tau_\varepsilon \mathcal{E}_{\phi\phi} - y^* \Psi''(P + c^* P')]$$

$$- \left. \frac{2 \tau_\varepsilon \mathcal{E}_\phi^2 \mathcal{E}_y}{y^*} \left(1 - \eta_{\mathcal{E}_\phi y} \right) \right] < 0 . \qquad (3.38)$$

The signs of these expressions are obtained from (3.30), $|J^*| > 0$, $\mathcal{X} < 0$, $\eta_{\mathcal{E}_\phi y} \leq 1$, $\eta_{\mathcal{E}_\phi} \leq \eta_{\Psi_\phi} \eta_{\mathcal{E}_y}$ and from $\Gamma'/\Gamma > \Psi'/\Psi$ in Lemma 2.1i on p. 16. Note that $\eta_{\mathcal{E}_\phi} \leq \eta_{\Psi_\phi} \eta_{\mathcal{E}_y}$ implies $\eta_{\mathcal{E}_\phi} \leq \eta_{\Gamma_\phi} \eta_{\mathcal{E}_y}$ since $\Gamma'/\Gamma > \Psi'/\Psi$. Using (3.37), (3.38) and $P' < 0$ in (3.29) proves $\tau_\varepsilon^{sb} < E'$.   □

Comparing Proposition 3.8 with Proposition 3.9 clarifies that, under procedure P1, we need an additional assumption for the second-best emissions tax to fall short of marginal damage. The emissions technology $\mathcal{E}$ has to be more output intensive in the sense that the elasticity of the emissions function with respect to durability, $\eta_{\mathcal{E}_\phi}$, is sufficiently smaller than the elasticity of the emissions function with respect to output, $\eta_{\mathcal{E}_y}$; note that η_{Ψ_ϕ} is less than one as long as Ψ is concave. Therefore, the set of assumptions listed in Proposition 3.9 is more restrictive than that listed in Proposition 3.8. Yet, an overinternalization case, if it exists, has counterintuitive implications also under procedure P1. To see this, define

$$\tilde{c} := y \Psi(\phi) . \qquad (3.39)$$

$\tilde{c}$ is the *approximate* long run stock of the durable good since it can be written as $\tilde{c} = y \Gamma(\phi) - y[\Gamma(\phi) - \Psi(\phi)] = c - y[\Gamma(\phi) - \Psi(\phi)]$. The approximation error

$\Delta := y[\Gamma(\phi) - \Psi(\phi)]$ is positive so that the approximate long run stock $\tilde{c}$ falls short of the exact long run stock c. Furthermore, the approximation error is the smaller, the smaller is the discount rate δ. With the help of this definition, under procedure P1 the second-best emissions tax (3.29) has to be written as

$$\tau_\varepsilon^{sb} = E' + c^* P' \frac{\partial \tilde{c}^* / \partial \tau_\varepsilon}{\partial \varepsilon^* / \partial \tau_\varepsilon} , \tag{3.40}$$

where $\tilde{c}^*$ is the approximate stock of the durable good in the long run market equilibrium. (3.40) is almost the same as (3.34) which reflects the second-best emissions tax in case that comparative statics of the market equilibrium are carried out with the help of procedure P2. The only difference is that (3.34) contains the influences of the emissions tax on the long run stock of the durable good whereas (3.40) contains the influence of the emissions tax on the *approximate* long run stock of the durable good. Consequently, we draw the same conclusion from (3.40) as from (3.34): Overinternalization to be second-best optimal requires that either equilibrium emissions or the equilibrium approximate stock of the durable good *increase* as the emissions tax increases. By the same arguments as presented for the long run stock in the discussion of (3.34), also a negative correlation between the emissions tax and the approximate stock seems to be plausible and, thus, the second-best emissions tax in G&B's polluting durable good industry (3.1)–(3.5) is expected to fall short of the marginal environmental damage independent of which procedure is employed in the comparative static analysis of the market equilibrium.

Of course, plausibility arguments are no perfect substitute for hard empirical evidence. Hence, although G&B are mistaken in identifying an overinternalization case and although they use not the exact procedure for the comparative static analysis, they are right in pointing out that overinternalization cannot be ruled out to be second-best optimal in a durable good industry, in general. However, owing to Proposition 3.8 and 3.9 a necessary condition for $\tau_\varepsilon^{sb} > E'$ is $\eta_{\mathcal{E}_\phi y} > 1$ which in turn implies $\mathcal{E}_{y\phi} > 0$. This observation is *not* an entirely new insight since it has already been derived by Barnett (1980) and Ebert (1992) in a similar way. In a nondurable good model, these authors assume costs and emissions functions similar to those used in the present durable good model, namely $K(y, \phi)$ and $\mathcal{E}(y, \phi)$ with $K_\phi > 0$ and $\mathcal{E}_\phi < 0$ where ϕ is interpreted not as product durability, but as pollution abatement effort of the individual firm. In their model, a necessary condition for overinternalization is $K_{y\phi} + \tau_\varepsilon \mathcal{E}_{y\phi} < 0$ which, in turn, implies $\mathcal{E}_{y\phi} < 0$ if the cost function is linear in output ($K_{y\phi} = \kappa' > 0$) as assumed in the present durable good model. Since the sign of $\mathcal{E}_\phi$ is reversed, this necessary condition is analogous to $\mathcal{E}_{y\phi} > 0$ in G&B's durable good model.

To sum up the discussion in this section, the possibility of overinternalization to be second-best optimal in a durable good industry (with precommitment of producers) cannot be excluded, in general. However, overinternalization (a) has counterintuitive implications, (b) is not really a new insight since it has already been derived in similar form by Barnett (1980) and Ebert

(1992) for nondurable goods and thus (c) is *not* the result of a 'third distortion' inherent in durable good markets, but is merely due to the special form of the emissions function.

3.2 Modified Model with Production Emissions

3.2.1 Model Description and First-Best Welfare Optimum

With respect to environmental policy, the main focus of G&B's analysis of the durable good industry is second-best emissions taxation. In G&B's model, one can hardly investigate other aspects of environmental policy: As shown in Proposition 3.4, setting the emissions tax rate equal to zero (condition C2) is not compatible with the existence of an interior solution of the firm's profit maximization (condition C1). Consequently, it is hardly possible to analyze the laissez faire allocation in the durable good industry or the effects of other policy instruments like e.g. an output tax.

To avoid these problems and to make the policy analysis richer, we now modify G&B's model. The modified model is still represented by equations (3.1)–(3.5). But rather than the linear cost function (3.1), we now consider the more general function

$$k_i(t) = K[y_i(t), \phi_i(t)] \, . \tag{3.41}$$

K in (3.41) is supposed to satisfy assumption A3 on p. 24. It covers the linear cost function as special case, but also allows for cost functions which are nonlinear in output. On the one hand, this modification makes the model more general and in some instances avoids the problems incurred in the previous section. On the other hand, however, the modification renders the analysis of the durable good industry more complicated. In order to keep the analysis tractable, the model is therefore simplified in another of its building blocks. More specifically, we replace the emissions function (3.4) by

$$\varepsilon_i(t) = \mathcal{E}[y_i(t)] \, . \tag{3.42}$$

The emissions function (3.42) is assumed to satisfy $\mathcal{E}'(y) > 0$ and $\mathcal{E}''(y) > 0$ for all $y \in \mathbb{R}_+$, i.e. firm i's emissions increase with increasing output at increasing rates. In contrast to G&B's model, firm i's emissions function does not depend on product durability. This is a reasonable assumption for durable goods whose durability can be enlarged without further material input. For example, it is often possible to prolong a product's lifetime by substituting a durable material for another, less durable one or by changing the product's design such that its rigidity is increased.

As a point of reference, we first characterize the efficient allocation in the modified durable good model. Imagine a social planer who maximizes social welfare. Seeking a symmetric solution $x_i(0, \infty) = x(0, \infty)$ for all $x \in \{y, \phi, c\}$ and all $i \in N$, the social planner solves the problem of

$$
\left.
\begin{aligned}
\max_{\substack{y(0, \infty) \\ \phi(0, \infty)}} V &= \int_0^\infty \left[S[nc(t)] - nK[y(t), \phi(t)] - E\left[n\mathcal{E}[y(t)]\right] \right] e^{-\delta t} dt \\
\text{subject to} \quad c(t) &= \int_{-\infty}^t D[t - v, \phi(v)] y(v) dv \ .
\end{aligned}
\right\} \quad (3.43)
$$

This maximization problem states that the social planer maximizes the present value V of social welfare over the planning period $[0, \infty[$ taking into account the integral constraint for the stock of the durable good. Instantaneous social welfare equals consumer benefits S less production costs nK and environmental costs E. (3.43) is an optimal control problem with integral state constraint. $y(t)$ and $\phi(t)$ are the control variables and $c(t)$ represents the state variable. For solving this problem, we introduce the present value Hamiltonian

$$
\mathcal{H}[y(t), \phi(t), c(t), \theta(t, \infty), t] = \left[S[nc(t)] - nK[y(t), \phi(t)] - E\left[n\mathcal{E}[y(t)]\right] \right] e^{-\delta t}
$$

$$
+ \int_t^\infty D[v - t, \phi(t)] y(t) \theta(v) dv \ ,
$$

where $\theta(t)$ is the costate variable or the social shadow price of the durable stock at time t. $\mathcal{H}$ measures the impact of time t control and state variables on the present value of social welfare. The state variable $c(t)$ at time t influences consumer benefits S at time t. The control variables $y(t)$ and $\phi(t)$ at time t have an effect on production and environmental costs at time t, K and E, and on the durable stock at $v \geq t$. The latter effect is represented by the integral in the Hamiltonian $\mathcal{H}$.

With the help of the Hamiltonian, we now apply the maximum principle derived by Bakke (1974) to the maximization of social welfare in (3.43). For simplicity, we focus on an interior solution for all variables and assume that the Hamiltonian is globally and strictly concave in the control variables. The maximum principle for problem (3.43) and the concavity property of the Hamiltonian then imply

$$
\theta(t) = \mathcal{H}_c(t) = nS'[nc^o(t)] e^{-\delta t} = nP[nc^o(t)] e^{-\delta t} \ , \qquad (3.44)
$$

$$
\mathcal{H}_y(t) = \mathcal{H}_\phi(t) = 0 \ , \qquad (3.45)
$$

$$
\mathcal{H}_{yy}(t) < 0 \ , \quad \mathcal{H}_{\phi\phi}(t) < 0 \ , \quad \mathcal{H}_{yy}(t)\mathcal{H}_{\phi\phi}(t) - \mathcal{H}_{y\phi}^2(t) > 0 \ , \qquad (3.46)
$$

for all $t \in [0, \infty[$. The superscript 'o' indicates efficient values. According to (3.44), the social shadow price of the durable good equals marginal consumer benefits from consuming the services of the durable good. Computing the derivatives in (3.45) and using (3.44) yields

$$K_y(t) + \mathcal{E}'(t)E'(t) = \int_t^\infty D(v)P(v)e^{-\delta(v-t)}dv , \qquad (3.47)$$

$$\frac{K_\phi(t)}{y^o(t)} = \int_t^\infty D_\phi(v)P(v)e^{-\delta(v-t)}dv , \qquad (3.48)$$

for all $t \in [0, \infty[$. The LHS of (3.47) represents marginal social costs of time t output. These costs consist of marginal production and environmental costs. The RHS of (3.47) gives marginal social benefits of time t output. These benefits reflect marginal consumer benefits of the durable stock at times $v \geq t$ since an increase in output at time t expands the durable stock at times $v \geq t$. In sum, (3.47) states that in the social optimum, marginal social costs of time t output equal marginal social benefits of time t output. Similar, (3.48) equates marginal social costs (LHS) and marginal social benefits (RHS) of time t product durability. Marginal social costs comprise marginal production costs only since in our modified model emissions depend on output, but not on durability. Marginal social benefits equal marginal consumer benefits accruing at times $v \geq t$ due to an increase in the durable stock.[7]

Since the emissions function (3.42) does not depend on product durability, there is no 'direct' relationship between production emissions and durability. This observation is confirmed by the marginal condition (3.48) in which marginal social costs of durability consist of marginal production costs only, but not of marginal environmental costs. Nevertheless, there is an 'indirect' impact of durability on emissions. This impact becomes best be seen by focusing on the long run welfare optimum. If we use definition D5 of an (asymptotic) stationary state and suppose that assumption A9 is satisfied, then conditions (3.47), (3.48) and the constraint for $c(t)$ in (3.43) become

$$K_y(y^o, \phi^o) + \mathcal{E}'(y^o)E'[n\mathcal{E}(y^o)] = P(nc^o)\Psi(\phi^o) , \qquad (3.49)$$

$$\frac{K_\phi(y^o, \phi^o)}{y^o} = P(nc^o)\Psi'(\phi^o) , \qquad (3.50)$$

$$c^o = y^o\Gamma(\phi^o) . \qquad (3.51)$$

These conditions determine long run efficient output y^o, product durability ϕ^o and durable stock c^o. As the general conditions (3.47) and (3.48), the long run conditions (3.49) and (3.50) equalize marginal social costs and benefits of output and product durability, respectively. Solving (3.49) for P and inserting the resulting expression into (3.50) gives

---

[7] The maximized Hamiltonian is concave in the state variable since $\mathcal{H}_{cc}^{max}(t) = n^2P'(t)e^{-\delta t} < 0$. Consequently, (3.47) and (3.48) together with the integral constraint in (3.43) are necessary and sufficient for the welfare optimum.

$$\frac{K_\phi(y^o,\phi^o)}{y^o} = K_y(y^o,\phi^o)\frac{\Psi'(\phi^o)}{\Psi(\phi^o)} + \mathcal{E}'(y^o)E'[n\mathcal{E}(y^o)]\frac{\Psi'(\phi^o)}{\Psi(\phi^o)} \; . \qquad (3.52)$$

(3.52) expresses in an alternative way the equality of marginal social costs and marginal social benefits of product durability. Marginal social costs on the LHS of (3.52) still equal marginal production costs. However, marginal social benefits on the RHS of (3.52) do not consist of marginal consumer benefits, but comprise two other marginal effects. To interpret these effects, remember that output and durability are substitutes with respect to the stock of the durable good. For constant stock, increasing product durability allows for reducing output and therefore lowers production costs. The saved production costs are represented by $K_y\Psi'/\Psi$ in (3.52). Decreasing output, in turn, implies that production emissions are reduced. Hence, for a constant stock of the durable good, enlarging product durability possesses an *emissions-reducing effect*. The associated savings in environmental costs are captured by $\mathcal{E}'E'\Psi'/\Psi$ in (3.52).

Conducting a comparative static analysis of the long run efficiency conditions (3.49)–(3.51) provides additional information about the properties of the welfare optimum. In comparison with previous durable good models already referred to in the introduction, the decisive differences of the model investigated in the present chapter is that it takes into account the environmental damage caused by the production of the durable good. Consequently, it is of particular interest to show how variations in marginal environmental costs alter the long run welfare optimum. For that purpose, introduce a shift parameter γ with $E'_\gamma > 0$. Increasing γ stands for an increase in marginal environmental costs. Totally differentiating (3.49)–(3.51) and applying Cramer's rule yields[8]

$$\frac{\partial y^o}{\partial \gamma} = -\frac{\mathcal{E}'E'_\gamma}{|J^o|}\left[K_{\phi\phi} - y^o\Psi''P - ny^{o2}\Gamma'\Psi'P'\right] , \qquad (3.53)$$

$$\frac{\partial \phi^o}{\partial \gamma} = \frac{\mathcal{E}'E'_\gamma}{|J^o|}\left[K_{y\phi} - \frac{K_\phi}{y^o} - ny^o\Gamma\Psi'P'\right] , \qquad (3.54)$$

$$\frac{\partial \varepsilon^o}{\partial \gamma} = \mathcal{E}'\frac{\partial y^o}{\partial \gamma} , \qquad (3.55)$$

$$\frac{\partial \tilde{c}^o}{\partial \gamma} = \frac{\mathcal{E}'E'_\gamma}{|J^o|}\left[y^o\Psi'\left(K_{y\phi} - \frac{K_\phi}{y^o}\right) - \Psi\left(K_{\phi\phi} - y^o\Psi''P\right)\right]$$

---

[8] From now on we will always apply procedure P1 in carrying out comparative statics. Furthermore, for later purpose, we compute not the comparative static results with respect to the long run stock $c = y\Gamma(\phi)$, but with respect to the approximate long run stock $\tilde{c} = y\Psi(\phi)$ defined in (3.39). This is done throughout the subsequent analysis. It can be shown that all results derived for $\tilde{c}$ hold also for c.

$$+ ny^{o2}\Psi'P'\left(\Gamma'\Psi - \Gamma\Psi'\right)\right]. \qquad (3.56)$$

$|J^o|$ is the Jacobian determinant of the long run efficiency conditions (3.49)–(3.51). The sign of $|J^o|$ is indeterminate, in general. However, it is possible to identify special cases in which $|J^o| > 0$, for example, the case in which the cost function is linear in output. We therefore assume $|J^o| > 0$ throughout. (3.53)–(3.56) then imply

Proposition 3.10. *In the long run social optimum*

$$\frac{\partial y^o}{\partial \gamma} < 0, \qquad \frac{\partial \phi^o}{\partial \gamma} > 0, \qquad \frac{\partial \varepsilon^o}{\partial \gamma} < 0,$$

$$\frac{\partial \tilde{c}^o}{\partial \gamma} < 0 \quad \text{if } K_{yy} = 0 \qquad \text{and} \qquad \frac{\partial \tilde{c}^o}{\partial \gamma} \gtreqless 0 \quad \text{otherwise}.$$

Proof. In the long run social optimum, the concavity condition $\mathcal{H}_{\phi\phi}(t) < 0$ in (3.46) is turned into $K_{\phi\phi} - y^o\Psi''P > 0$. Furthermore, the Abel condition in assumption A3 implies $K_{y\phi} - K_\phi/y^o = K_{yy}K_\phi/K_y \geq 0$. $P' < 0$ and (3.53) - (3.55) then prove $\partial y^o/\partial\gamma < 0$, $\partial\phi^o/\partial\gamma > 0$ and $\partial\varepsilon^o/\partial\gamma < 0$. The sign of $\partial\tilde{c}^o/\partial\gamma$ in (3.56) is indeterminate, in general. However, in the special case of $K_{yy} = 0$, we obtain $K_{y\phi} - K_\phi/y^o = 0$. In view of this equation and $\Gamma'/\Gamma > \Psi'/\Psi$ from Lemma 2.1, we have $\partial\tilde{c}^o/\partial\gamma < 0$.   □

The intuition of Proposition 3.10 is as follows. Increasing γ raises marginal environmental costs of emissions. Consequently, the social planer reduces the efficient long run amount of emissions. This is done by lowering output of the durable good and by taking advantage of the emissions-reducing effect of increasing product durability. Hence, durability is raised. It is generally not clear whether a parametric increase in marginal environmental damage increases or decreases the efficient long run stock of the durable good since this stock is positively correlated with efficient long run output which decreases and efficient long run durability which increases. However, in the special case of a linear cost function, the decreasing output overcompensates the enlarged durability so that the efficient long run stock is unambiguously reduced when the marginal environmental damage becomes greater.

3.2.2 Market Equilibrium and Laissez Faire

Having investigated the social welfare optimum we now turn to the market equilibrium. It is again assumed that the government levies a tax on production emissions. In contrast to G&B's model, however, we suppose that the associated tax rate $\tau_\varepsilon(t)$ depends on time since we are interested in environmental regulation not only in the long run, but also on the adjustment path of the durable good economy. To enrich the policy analysis further, it is supposed that the government regulates output, product durability and the stock of the durable good, too. $\tau_y(t)$, $\tau_\phi(t)$ and $\tau_c(t)$ are the respective tax rates at time t. We adopt the convention that a tax rate represents a tax (subsidy) when it is positive (negative). Taking into account these tax rates, firm i's instantaneous profits equal rental revenues less production costs and tax payments for emissions, output, product durability and the stock of the durable good. The pay-off of firm i is a function $\Pi^{ri} : S^n \mapsto \mathbb{R}$ where

$$\Pi^{ri}(s_i, s_{-i}) = \int_0^\infty \left[P(\cdot) \int_{-\infty}^t D[t - v, \phi_i(v)] y_i(v) dv - K[y_i(t), \phi_i(t)] \right.$$

$$- \tau_\varepsilon(t)\mathcal{E}[y_i(t)] - \tau_y(t)y_i(t) - \tau_c(t) \int_{-\infty}^t D[t - v, \phi_i(v)] y_i(v) dv$$

$$\left. - \tau_\phi(t)\phi_i(t) \right] e^{-\delta t} dt \qquad (3.57)$$

is the present value of firm i's rental profits over the planning period $[0, \infty[$. We again assume an open-loop information pattern and derive the open-loop rental equilibrium defined by definition D2. As in Proposition 2.2 on p. 33, it can be shown that the open-loop sales equilibrium is identical to the open-loop rental equilibrium if assumption A5 or assumption A6 is satisfied. Note that this equivalence is not trivial in the presence of a tax on the durable stock. This tax is paid by the owners of the durable goods who in sales markets are consumers and in rental markets are producers. So, in the rental case, the tax on the durable stock tax enters the profit functions of the durable good firms while in the sales case, the tax on the durable stock enters the utility maximization calculus of consumers. However, utility-maximizing consumers adjust their willingness-to-pay for the durable good by the tax on the durable stock with the consequence that $\tau_c(t)$ enters the firms' profit functions in the sales case, too. The same argument holds if output and durability taxes are levied not on the firms' supply, but on the consumers' demand of the durable good.[9]

Before proceeding with the characterization of the market equilibrium, it should be noted that we are interested in both perfect and imperfect competition. Under imperfect competition, firms perceive their impact on the price of the durable's services, i.e. they take $P' < 0$ into account. Under perfect competition, firms take the price of the durable's service as exogenously given. This price taking behavior is captured in the subsequent analysis if we set $P' \equiv 0$.

Following definition D2, the durable good industry attains an equilibrium if all firms mutually give the best response to the behavior of the respective

---

[9] These assertions become obvious in Chap. 5 where we consider utility maximization of a representative consumer. See especially footnote 7 on p. 153.

competitors. Hence, firm i chooses its strategy s_i in order to maximize its pay-off function $\Pi^{ri}(s_i, s_{-i})$ taking as given the strategies s_{-i} of its competitors. This maximization problem can be written as

$$
\max_{\substack{y_i(0,\infty) \\ \phi_i(0,\infty)}} \int_0^\infty \Bigg[c_i(t)P[c_i(t) + C^i(t)] - K[y_i(t), \phi_i(t)] - \tau_\varepsilon(t)\mathcal{E}[y_i(t)]
$$

$$
- \tau_y(t)y_i(t) - \tau_c(t)c_i(t) - \tau_\phi(t)\phi_i(t) \Bigg] e^{-\delta t} dt \quad (3.58)
$$

$$
\text{subject to} \quad c_i(t) = \int_{-\infty}^t D[t - v, \phi_i(v)]y_i(v)dv .
$$

In order to characterize the solution to this optimal control problem with integral state constraint, define firm i's Hamiltonian as

$$
\mathcal{H}[y_i(t), \phi_i(t), c_i(t), \lambda_i(t, \infty), t] = \Big[c_i(t)P[C(t)] - K[y_i(t), \phi_i(t)] - \tau_\varepsilon(t)\mathcal{E}[y_i(t)]
$$

$$
- \tau_y(t)y_i(t) - \tau_c(t)c_i(t) - \tau_\phi(t)\phi_i(t) \Big] e^{-\delta t}
$$

$$
+ \int_t^\infty D\big[v - t, \phi_i(t)\big]y_i(t)\lambda_i(v)dv ,
$$

where $\lambda_i(t)$ is firm i's shadow price of the durable stock at time t. We again assume an interior solution and a strictly concave Hamiltonian. The maximum principle and the concavity of the Hamiltonian then yield

$$
\lambda_i(t) = \mathcal{H}_c(t) = \Big[P[C(t)] + c_i^*(t)P'[C(t)] - \tau_c(t) \Big] e^{-\delta t} , \quad (3.59)
$$

$$
\mathcal{H}_y(t) = \mathcal{H}_\phi(t) = 0 , \quad (3.60)
$$

$$
\mathcal{H}_{yy}(t) < 0, \quad \mathcal{H}_{\phi\phi}(t) < 0 , \quad \mathcal{H}_{yy}(t)\mathcal{H}_{\phi\phi}(t) - \mathcal{H}_{y\phi}^2(t) > 0 , \quad (3.61)
$$

for all $t \in [0, \infty[$. Note that in the special case in which the cost function is linear ($K_{yy} = 0$) and the emissions tax rate is zero ($\tau_\varepsilon = 0$), we obtain condition C2 from G&B's model presented in Sect. 3.1. According to Proposition 3.4, this condition is not compatible with condition C1, i.e. in the modified model it is not compatible with an interior solution determined by (3.59)–(3.61). To avoid this problem, the subsequent analysis implicitly assumes $\tau_\varepsilon > 0$ ($K_{yy} > 0$) whenever the special case $K_{yy} = 0$ ($\tau_\varepsilon = 0$) is considered.

According to (3.59), firm i's shadow price of the durable good at time t equals marginal revenues of the durable stock less stock tax payments at time t. Computing (3.60) and invoking (3.59) gives

$$
K_y(t) + \tau_\varepsilon(t)\mathcal{E}'(t) + \tau_y(t) + \int_t^\infty D(v)\tau_c(v)e^{-\delta(v-t)}dv
$$

$$= \int_t^\infty D(v)[P(v) + c_i^*(v)P'(v)]e^{-\delta(v-t)}dv , \quad (3.62)$$

$$\frac{K_\phi(t) + \tau_\phi(t)}{y_i^*(t)} + \int_t^\infty D_\phi(v)\tau_c(v)e^{-\delta(v-t)}dv$$

$$= \int_t^\infty D_\phi(v)[P(v) + c_i^*(v)P'(v)]e^{-\delta(v-t)}dv , \quad (3.63)$$

for all $t \in [0, \infty[$. In order to attain a profit maximum, firm i has to set time t output and product durability such that marginal costs of these variables coincide with marginal revenues. As in G&B's model, marginal revenues reflect marginal revenues of the durable stock at times $v \geq t$. According to the LHS of (3.62), however, marginal costs of time t output equal not only the sum of marginal production costs and marginal emissions tax payments, but also comprise marginal tax payments for output and the stock of the durable good. Marginal tax payments for the durable stock arise at times $v \geq t$ since increasing output at t enlarges the durable stock at times $v \geq t$. The same is true for marginal costs of time t durability on the LHS of (3.63). These marginal costs reflect the respective marginal production costs at time t plus marginal tax payments for durability at time t and marginal tax payments for the stock of the durable good at times $v \geq t$.[10]

Conditions for a symmetric long run market equilibrium are obtained from (3.62), (3.63) and the integral constraint in (3.58) if we presuppose definition D5 and assumption A9. The long run conditions then read

$$K_y(y^*, \phi^*) + \tau_\varepsilon \mathcal{E}(y^*) + \tau_y + \tau_c \Psi(\phi^*) = [P(nc^*) + c^* P'(nc^*)]\Psi(\phi^*) , \quad (3.64)$$

$$\frac{K_\phi(y^*, \phi^*) + \tau_\phi}{y^*} + \tau_c \Psi'(\phi^*) = [P(nc^*) + c^* P'(nc^*)]\Psi'(\phi^*) , \quad (3.65)$$

$$c^* = y^* \Gamma(\phi^*) , \quad (3.66)$$

where y^*, ϕ^* and c^* denote long run equilibrium values of output, product durability and stock of the durable good, respectively. Totally differentiating (3.64)–(3.66) and applying Cramer's rule yields

$$\frac{\partial \phi^*}{\partial n} = \frac{c^*(P' + c^* P'')}{|J^*|}\left[\frac{\tau_\phi K_{yy}}{P + c^* P' - \tau_c} - \frac{\tau_\varepsilon \mathcal{E}' + \tau_y}{P + c^* P' - \tau_c}\frac{K_\phi K_{yy}}{K_y}\right.$$

$$\left. + \tau_\varepsilon y^* \Psi' \mathcal{E}_{yy} + \frac{\tau_\phi \Psi}{y^*}\right] . \quad (3.67)$$

---

[10] Note that (3.62) and (3.63) jointly with the integral constraint in (3.58) are not only necessary, but also sufficient for the solution of (3.58) since the maximized Hamiltonian is concave in the stock of the durable good according to $\mathcal{H}_{cc}^{\max}(t) = [2P'(t) + c_i^*(t)P''(t)]e^{-\delta t} < 0$.

$|J^*|$ is the Jacobian determinant of the long run equilibrium conditions (3.64)–
(3.66). In general, $|J^*|$ may take on any sign. However, it is possible to secure
$|J^*| > 0$ at least in some special cases, for example, if the cost function is
linear in output ($K_{yy} = 0$) and the effluent charge is strictly positive ($\tau_\varepsilon > 0$).
Moreover, in footnote 12 on p. 82, the condition $|J^*| > 0$ is shown to be
necessary and sufficient for the existence of a second-best emissions tax. We
therefore assume $|J^*| > 0$ throughout. From (3.67) immediately follows

Proposition 3.11. *In the long run market equilibrium*

(i) $\quad \dfrac{\partial \phi^*}{\partial n} = 0 \quad if \quad \tau_\varepsilon = \tau_y = \tau_\phi = 0 \,,$

(ii) $\quad \dfrac{\partial \phi^*}{\partial n} \neq 0 \quad if \quad \tau_\varepsilon \neq 0 \quad or \quad \tau_y \neq 0 \quad or \quad \tau_\phi \neq 0 \,.$

According to this proposition, the validity of Swan's independence result
depends on the emissions tax rate, the output tax rate and the durability
tax rate. The rationale is as follows. Firms set product durability in a cost-
minimizing way. If the tax rates on emissions, output and product durability
are equal to zero, then costs consist of production costs only (when we ignore, for the time being, the tax payments for the stock of the durable good).
Due to the Abel condition, minimization of production costs is independent
of the firms' market share and, hence, cost-minimizing product durability is
independent of market structure as shown in Proposition 3.11i. In contrast,
if at least one of the tax rates on emissions, output and durability is different from zero, then the firms' costs comprise not only production costs,
but also tax payments. Consequently, the Abel condition is not sufficient to
render cost minimization independent of the firms' market share and, thus,
cost-minimizing product durability is influenced by the market structure as
spelled out in Proposition 3.11ii.

One may wonder why Proposition 3.11 does not include the tax payments
for the stock of the durable good. The reason is that these tax payments are
effectively equivalent to a change in the price of the durable's services. The
price is reduced if τ_c is positive whereas a negative τ_c increases the price.
Hence, the stock tax influences the demand conditions, but leaves unaltered
cost minimization and the cost-minimizing product durability. The stock tax
is therefore irrelevant for the validity of Swan's independence result. Formally, this is expressed in Proposition 3.11i which implies that Swan's result
does hold when the stock tax rate is the only non-zero policy instrument. In
comparison to G&B's model, the analysis of the market equilibrium in the
modified model therefore provides a more comprehensive view on the relation
between taxation and the Swan's independence result. On the one hand, it
shows that emissions taxes are not the only taxes which upset Swan's result.
On the other hand, the modified model reveals that for some types of taxes,
the independence of product durability and market structure remains valid.

For later reference, it is also interesting to know the impact of emissions
and output taxation on the long run market equilibrium. To simplify, we

set the other tax rates equal to zero ($\tau_c = \tau_\phi = 0$). For $\tau \in \{\tau_\varepsilon, \tau_y\}$, the comparative static analysis then yields

$$\frac{\partial y^*}{\partial \tau} = -\frac{\mathcal{R}}{|J^*|} \left[K_{\phi\phi} - y^* \Psi''(P + c^* P') - y^{*2} \Gamma' \Psi' \mathcal{X} \right], \tag{3.68}$$

$$\frac{\partial \phi^*}{\partial \tau} = \frac{\mathcal{R}}{|J^*|} \left[K_{y\phi} - \frac{K_\phi}{y^*} - y^* \Gamma \Psi' \mathcal{X} \right], \tag{3.69}$$

$$\frac{\partial \varepsilon^*}{\partial \tau} = \mathcal{E}' \frac{\partial y^*}{\partial \tau}, \tag{3.70}$$

$$\frac{\partial \tilde{c}^*}{\partial \tau} = \frac{\mathcal{R}}{|J^*|} \left[y^* \Psi' \left(K_{y\phi} - \frac{K_\phi}{y^*} \right) - \Psi \left(K_{\phi\phi} - y^* \Psi''(P + c^* P') \right) \right.$$
$$\left. + y^{*2} \Psi' \mathcal{X} \left(\Gamma' \Psi - \Gamma \Psi' \right) \right], \tag{3.71}$$

where $\mathcal{R} = \mathcal{E}'$ if $\tau = \tau_\varepsilon$ and $\mathcal{R} = 1$ if $\tau = \tau_y$. $\tilde{c}^* = y^* \Psi(\phi^*)$ is the approximate stock of the durable good in the long run equilibrium. Under imperfect competition, P' in (3.64)–(3.66) is negative and, thus, $\mathcal{X}$ in (3.68)–(3.71) equals $(n+1)P' + nc^* P''$. Under perfect competition, P' in (3.64)–(3.66) is identical to zero and $\mathcal{X}$ in (3.68)–(3.71) simplifies to nP'. In both cases, we obtain $\mathcal{X} < 0$ owing to assumption A4. From (3.68)–(3.71) then follows

Proposition 3.12. *Set $\tau_c = \tau_\phi = 0$ and let $\tau \in \{\tau_\varepsilon, \tau_y\}$. Then, in the long run market equilibrium*

$$\frac{\partial y^*}{\partial \tau} < 0, \qquad \frac{\partial \phi^*}{\partial \tau} > 0, \qquad \frac{\partial \varepsilon^*}{\partial \tau} < 0,$$

$$\frac{\partial \tilde{c}^*}{\partial \tau} < 0 \quad \textit{if } K_{yy} = 0 \qquad \textit{and} \qquad \frac{\partial \tilde{c}^*}{\partial \tau} \gtreqless 0 \quad \textit{otherwise}.$$

These results hold under both perfect and imperfect competition.

Proof. In the long run, the concavity condition $\mathcal{H}_{\phi\phi}(t) > 0$ in (3.61) becomes $K_{\phi\phi} - y^* \Psi''(P + c^* P') > 0$. Furthermore, the Abel condition implies $K_{y\phi} - K_\phi/y^* = K_{yy} K_\phi / K_y > 0$. $\mathcal{X} < 0$, $\mathcal{R} > 0$, $|J^*| > 0$ and (3.68)–(3.70) then prove $\partial y^*/\partial \tau < 0$, $\partial \phi^*/\partial \tau > 0$ and $\partial \varepsilon^*/\partial \tau < 0$. The sign of $\partial \tilde{c}^*/\partial \tau$ is indeterminate, in general. However, for $K_{yy} = 0$, we obtain $K_{y\phi} - K_\phi/y^* = 0$ and $\partial \tilde{c}^*/\partial \tau < 0$. $\square$

To interpret Proposition 3.12, consider first the case of emissions taxation ($\tau = \tau_\varepsilon$). An increase in the emissions tax rate raises the costs of polluting the environment during the production of the durable good. Hence, no matter whether firms operate under perfect or imperfect competition, they reduce their emissions by producing fewer units of the durable good. Furthermore, they raise product durability since expanding the lifetime of the products allows reducing output, production emissions and the emissions tax bill. In general, it is not clear whether firms also reduce the stock of the durable good. However, in case of a linear cost function ($K_{yy} = 0$), the reduction in output outweighs the increase in product durability so that the equilibrium stock of the durable good declines. The interpretation regarding output taxation ($\tau = \tau_y$) is similar. Increasing the output tax provides firms with the incentive to substitute product durability for output with the consequence that production emissions decline. In case of a linear cost function, the effect on the stock of the durable good is negative.

Before turning to normative aspects of regulation in our polluting durable good industry, it is appropriate to make the case for regulation in the first place. This is typically done by demonstrating that markets fail under laissez faire. Let us therefore focus on the long run, set all tax rates equal to zero and compare the market equilibrium with the social welfare optimum. This yields

Proposition 3.13. *Set $\tau_\varepsilon = \tau_y = \tau_c = \tau_\phi = 0$. Then,*
(i) the long run market equilibrium under perfect competition satisfies

$$y^* > y^o, \qquad \phi^* < \phi^o, \qquad \varepsilon^* > \varepsilon^o, \qquad \tilde{c}^* \gtreqless \tilde{c}^o,$$

(ii) the long run market equilibrium under imperfect competition satisfies

$$y^* \gtreqless y^o, \qquad \phi^* < \phi^o, \qquad \varepsilon^* \gtreqless \varepsilon^o, \qquad \tilde{c}^* \gtreqless \tilde{c}^o,$$

where ϕ^* is the same as in the long run market equilibrium under perfect competition.

Proof. From Proposition 3.12 we know that $\partial y^*/\partial \tau_\varepsilon < 0$, $\partial \phi^*/\partial \tau_\varepsilon > 0$, $\partial \varepsilon^*/\partial \tau_\varepsilon < 0$ and $\partial \tilde{c}^*/\partial \tau_\varepsilon \gtreqless 0$, in general. Furthermore, in Table 3.1 of the next section, it is shown that under perfect competition $y^* = y^o$, $\phi^* = \phi^o$, $\varepsilon^* = \varepsilon^o$ and $\tilde{c}^* = \tilde{c}^o$ if $\tau_\varepsilon = E' > 0$ and $\tau_y = \tau_\phi = \tau_c = 0$. Consequently, for $\tau_\varepsilon = \tau_y = \tau_c = \tau_\phi = 0$, it follows $y^* > y^o$, $\phi^* < \phi^o$, $\varepsilon^* > \varepsilon^o$ and $\tilde{c}^* \gtreqless \tilde{c}^o$. This proves Proposition 3.13i. $\phi^* < \phi^o$ in Proposition 3.13ii follows directly from Swan's independence result which holds in the laissez faire economy according to Proposition 3.11i. According to Table 3.1 in the next section, under imperfect competition, we have $y^* = y^o$, $\varepsilon^* = \varepsilon^o$ and $\tilde{c}^* = \tilde{c}^o$ if $\tau_\varepsilon = E' > 0$, $\tau_c = c^o P' < 0$ and $\tau_y = \tau_\phi = 0$. Furthermore, it can be shown that $\partial y^*/\partial \tau_c \gtreqless 0$. This completes the proof of Proposition 3.13ii. □

Increasing the output of the durable good causes social costs since it raises production emissions so that environmental costs increase. In case of laissez faire under perfect competition considered in Proposition 3.13i, producers ignore this effect since environmental costs are external to them. Therefore, firms choose output and production emissions inefficiently high. Moreover, producers ignore the emissions-reducing effect of increasing durability. Hence, unregulated durable good producers set product durability below the efficient level. Inefficiently low product durability and inefficiently high output imply that the stock of the durable good may be inefficiently high or low in the laissez faire market equilibrium.[11] In the long run under imperfect competition considered in Proposition 3.13ii, there is a second market imperfection present, namely the market power of producers. This imperfection distorts output and production emissions into the other direction as the environmental externality and, hence, output and production emissions are not necessarily inefficiently high, but possibly attain inefficiently low levels. In contrast, product durability keeps falling short of its efficient level. The reason is Swan's independence result which ensures that product durability under imperfect competition takes on the same inefficiently low value as under perfect competition. This case of underprovision of product durability (irrespective of whether it occurs under perfect or imperfect competition) has not yet been established in the literature since the models studied so far either ignore environmental aspects of durable goods or consider nondurable goods. See the articles referred to in the introduction. G&B's approach presented in Sect. 3.1 is not suitable to establish this underprovision result, either, because of the technical problems associated with it.

3.2.3 First-Best Regulation

The next step is to investigate how the market failure identified in Proposition 3.13 can be corrected with the help of tax instruments. It is very plausible that the market failure occurs not only in the long run, but also on the adjustment path. Efficiency restoring regulatory schemes for the adjustment path are obtained by comparing the efficiency conditions (3.47) and (3.48) with the market equilibrium conditions (3.62) and (3.63). It follows that the market equilibrium exhibits an efficient allocation if and only if

$$\left[\tau_\varepsilon(t) - E'(t)\right]\mathcal{E}'(t) + \tau_y(t) = \int_t^\infty D(v)[c^o(v)P'(v) - \tau_c(v)]e^{-\delta(v-t)}dv, \quad (3.72)$$

$$\frac{\tau_\phi(t)}{y^o(t)} = \int_t^\infty D_\phi(v)[c^o(v)P'(v) - \tau_c(v)]e^{-\delta(v-t)}dv, \quad (3.73)$$

for all $t \in [0, \infty[$. According to (3.72) and (3.73), the most obvious regulation under perfect competition ($P' \equiv 0$) is to set $\tau_\varepsilon(t) = E'(t) > 0$ and $\tau_y(t) = \tau_c(t) = \tau_\phi(t) = 0$ for all $t \in [0, \infty[$. This is the conventional Pigouvian

---

[11] It is not possible to establish that the laissez faire stock of the durable good is inefficiently low if the cost function is linear in output: For $K_{yy} = \tau_e = 0$, we obtain condition C2 which is not compatible to our assumption of an interior solution. See Proposition 3.4.

approach which requires setting the emissions tax rate equal to marginal environmental costs. The emissions tax internalizes the environmental externality since producers take into account the marginal environmental damage caused by the production of the durable good. They reduce output and the associated production emissions, account for the emissions-reducing effect of durability and render their products more durable. Hence, the regulator does not need an additional instrument to correct for the distortion of product durability. According to (3.72) and (3.73), the most obvious regulation under imperfect competition $(P' < 0)$ is to combine the Pigouvian emissions tax with a subsidy on the stock of the durable good, i.e. $\tau_\varepsilon(t) = E'(t) > 0$, $\tau_c(t) = c^o(t)P'(t) < 0$ and $\tau_y(t) = \tau_\phi(t) = 0$ for all $t \in [0,\infty[$. The Pigouvian tax eliminates the environmental distortion, but leaves the durable stock inefficiently low owing to the market power of firms. The stock subsidy then provides firms with the incentive to shift the durable stock up to its efficient level.

In addition to these conventional regulatory schemes, there are other combinations of tax rates which render the market equilibrium socially optimal. To analyze additional policies, we focus on the long run. From (3.72) and (3.73) immediately follows

Proposition 3.14. *The long run market equilibrium is socially optimal if and only if*

$$\left[\tau_\varepsilon - E'[n\mathcal{E}(y^o)]\right]\mathcal{E}'(y^o) + \tau_y = \Psi(\phi^o)\left[c^o P'(nc^o) - \tau_c\right],$$

$$\tau_\phi = y^o \Psi'(\phi^o)\left[c^o P'(nc^o) - \tau_c\right].$$

According to this proposition, the regulator may implement convex combinations of the different tax rates in order to attain the first-best welfare optimum. Hence, the number of efficiency restoring policy schemes is infinite. This is not surprising since the regulator has at her disposal more tax instruments than distortions are present in the industry. She may set four tax rates (on emissions, output, durability and the stock of the durable good) while the market outcome is distorted only by two market imperfections, namely the environmental externality and the market power of producers.

Of course, the variety in the first-best regulatory schemes suggested by Proposition 3.14 is not very useful from a political point of view. Table 3.1 therefore derives from Proposition 3.14 all regulatory schemes consisting of no more than two non-zero tax rates. Consider first the case of perfect competition. Row 1 of Table 3.1 contains the Pigouvian scheme which we have already discussed above. According to row 2 of Table 3.1, it is also possible to tax output rather than emissions. The reason is that output determines production emissions by the emissions function $\mathcal{E}$ and, thus, the tax may be levied on the emissions generating activity rather than on the emissions themselves. Moreover, convex combinations of emissions and output taxes render the market equilibrium first-best provided the tax rates are set as specified in

Table 3.1. First-Best Regulatory Schemes in the Presence of Production Emissions

row	τ_ε	τ_y	τ_c	τ_ϕ
		Perfect competition		
1	$E' > 0$	–	–	–
2	–	$\mathcal{E}'E' > 0$	–	–
3	$\tau_\varepsilon = E' - \tau_y/\mathcal{E}'$		–	–
4	–	–	$\mathcal{E}'E'/\Psi > 0$	$-y^\circ\mathcal{E}'E'\Psi'/\Psi < 0$
		Imperfect competition		
5	$E' > 0$	–	$c^\circ P' < 0$	–
6	–	$\mathcal{E}'E' > 0$	$c^\circ P' < 0$	–
7	$E' + c^\circ \Psi P'/\mathcal{E}' \gtreqless 0$	–	–	$y^\circ c^\circ \Psi' P' < 0$
8	–	$\mathcal{E}'E' + c^\circ \Psi P' \gtreqless 0$	–	$y^\circ c^\circ \Psi' P' < 0$
9	–	–	$c^\circ P' + \mathcal{E}'E'/\Psi \gtreqless 0$	$-y^\circ\mathcal{E}'E'\Psi'/\Psi < 0$

row 3 of Table 3.1. Hence, the tax burden can be divided between emissions and output. Finally, if the regulator wants or needs to avoid output and emissions taxation, then she has the option to subsidize product durability and tax the stock of the durable good as shown in row 4 of Table 3.1. The subsidy on product durability induces producers to expand the lifetime of their products. This has the desired emissions-reducing effect provided the stock of the durable good remains unaltered. However, the increase in product durability exerts a positive effect on the durable stock. To neutralize this effect, the regulator needs to impose an additional stock tax which provides firms with the incentive for keeping the durable stock constant. Note that it is not feasible to implement the first-best optimum by exclusive durability regulation or by exclusive regulation of the durable stock.

According to rows 5–9 of Table 3.1, efficiency under imperfect competition cannot be restored by means of a single tax. For example, the emissions tax from row 1 or the output tax from row 2 of Table 3.1 has to be supplemented by a subsidy on the stock of the durable good. The emissions tax and the output tax, respectively, internalize the environmental externality of production emissions, and the stock subsidy corrects for the underprovision of the durable stock which is due to the market power of firms. Since the durable stock is determined by output and product durability, rows 7 and 8 of Table 3.1 show that the underprovision of the durable stock can also be avoided by subsidizing product durability and lowering the tax rate which corrects for the environmental externality. Furthermore, also under imperfect competition it is possible to implement the first-best allocation by subsidizing product durability and taxing the stock of the durable good. This is shown in row 9 of Table 3.1. In contrast to the case of perfect competition, however, the tax rate on the durable stock needs to account not only for the environmental externality, but also for the market power of producers and, thus, it is no longer clear whether the stock is subject to a tax as under perfect competition or whether the stock tax turns into a stock subsidy. Finally, note that a convex combination of emissions and output taxes is not an appropriate scheme to restore efficiency under imperfect competition (in contrast to perfect competition) since such a scheme ignores the market power distortion.

3.2.4 Second-Best Taxation

According to Table 3.1, the regulator is able to restore efficiency under perfect competition with a single tax instrument, but implementing the welfare optimum under imperfect competition requires at least two non-zero tax rates. If the regulator does not have at her disposal the appropriate instruments, then the market equilibrium deviates from the first-best optimum. This section conducts a second-best analysis for such situations. In contrast to G&B's analysis, we consider not only second-best emissions taxation, but also second-best output taxation.

For simplicity, we focus on the long run. The second-best emissions tax is then obtained by setting $\tau_y = \tau_c = \tau_\phi = 0$ and solving the welfare maximization problem

$$\left. \begin{aligned} \max_{\tau_\varepsilon} \ V(\tau_\varepsilon) \ &= \ S[nc^*(\tau_\varepsilon)] - nK[y^*(\tau_\varepsilon), \phi^*(\tau_\varepsilon)] - E\Big[n\mathcal{E}[y^*(\tau_\varepsilon)]\Big] \\ \text{subject to} \quad c^*(\tau_\varepsilon) &= y^*(\tau_\varepsilon)\Psi[\phi^*(\tau_\varepsilon)] - \nu_1\Psi(\nu_2) + \nu_1\Gamma(\nu_2) \, , \end{aligned} \right\} \quad (3.74)$$

where we account for the adjustment path by taking ν_1 and ν_2 as given. (3.74) is similar to the second-best welfare maximization (3.27) in G&B's model. The difference is that (3.74) employs the general cost function $K(y, \phi)$ rather than the linear cost function $\kappa(\phi)y$ and the simplified emissions function $\mathcal{E}(y)$ rather than the general emissions function $\mathcal{E}(y, \phi)$. Differentiating social welfare with respect to the emissions tax rate τ_ε and taking into account the market equilibrium conditions (3.64) and (3.65) yields

$$V'(\tau_\varepsilon) \gtreqqless 0 \qquad \Leftrightarrow \qquad \tau_\varepsilon \lesseqqgtr E' + c^*P'\frac{\partial \tilde{c}^*/\partial \tau_\varepsilon}{\partial \varepsilon^*/\partial \tau_\varepsilon} \, , \qquad (3.75)$$

where $\partial \varepsilon^*/\partial \tau_\varepsilon$ and $\partial \tilde{c}^*/\partial \tau_\varepsilon$ represent the changes in equilibrium emissions and the equilibrium (approximate) stock of the durable good due to changes in the emissions tax rate. These derivatives are specified in (3.70) and (3.71). (3.75) implies that the welfare function V is inverted U-shaped in the emissions tax rate. The welfare increases in τ_ε for all τ_ε smaller than the threshold value $E' + c^* P'(\partial \tilde{c}^*/\partial \tau_\varepsilon)/(\partial \varepsilon^*/\partial \tau_\varepsilon)$ and decreases for all τ_ε greater than the threshold value. Hence, the second-best emissions tax reads[12]

$$\tau_\varepsilon^{sb} = E' + c^* P' \frac{\partial \tilde{c}^*/\partial \tau_\varepsilon}{\partial \varepsilon^*/\partial \tau_\varepsilon} \,. \tag{3.76}$$

Taking into account the comparative static results of Proposition 3.12, equation (3.76) immediately implies

Proposition 3.15. *In the long run, the second-best emissions tax satisfies*

$$(i) \quad \tau_\varepsilon^{sb} < E' \quad if \ K_{yy} = 0 \qquad and \qquad (ii) \quad \tau_\varepsilon^{sb} \gtreqless E' \quad otherwise\,.$$

In the modified model of the present section, we have assumed that the emissions function is independent of product durability. Hence, the emissions function satisfies the conditions $\eta_{\mathcal{E}_\phi y} = 0 < 1$ and $\eta_{\mathcal{E}\phi} = 0 < \eta_{\Psi\phi}\eta_{\mathcal{E}y}$ which Proposition 3.9 identifies as sufficient for the second-best emissions tax to fall short of the marginal damage in G&B's model presented in Sect. 3.1. But these conditions turn out to be not sufficient for underinternalization in the modified model since this model assumes a general production cost function rather than a linear one. Of course, in case of a linear cost function, the second-best emissions tax is smaller than the marginal environmental costs in the modified model, too. This assertion is formally proved in Proposition 3.15i. Admittedly, assuming linear production costs is rather restrictive and overinternalization can therefore not be excluded as shown by Proposition 3.15ii. But similar to the discussion of (3.40) in G&B's model, an overinternalization case (if it exists) has counterintuitive implications since according to (3.76) it requires the equilibrium stock of the durable good to *rise* when the emissions tax rate is increased.

For investigating second-best output taxation, set $\tau_\varepsilon = \tau_c = \tau_\phi = 0$. Welfare maximization is then again represented by (3.74) except that τ_ε is replaced by τ_y. Differentiating social welfare V with respect to the output tax and using the market equilibrium conditions (3.64) and (3.65) gives

$$V'(\tau_y) \gtreqless 0 \qquad \Leftrightarrow \qquad \tau_y \lesseqgtr \mathcal{E}'E' + c^* P' \frac{\partial \tilde{c}^*/\partial \tau_y}{\partial y^*/\partial \tau_y} \,. \tag{3.77}$$

---

[12] $|J^*| > 0$ is a necessary and sufficient condition for the existence of a second-best emissions tax rate. Necessity can be proved by contradiction: Suppose a second-best emissions tax exists and $|J^*| < 0$. This implies the signs of the partial derivatives in Proposition 3.12 to be reversed such that $\partial \varepsilon^*/\partial \tau_\varepsilon > 0$. Hence, the inequality signs in the second relation of (3.75) are reversed, too. The welfare function is then U-shaped in τ_ε, and there does not exist an emissions tax which maximizes social welfare. This contradicts the initial assumption of the existence of a second-best emissions tax. Sufficiency is also straightforward in view of (3.75). Recall from p. 75 that we assumed $|J^*| > 0$ throughout.

$\partial y^*/\partial \tau_y$ and $\partial \tilde{c}^*/\partial \tau_y$ show how the equilibrium output and stock of the durable good change if the output tax rate is varied. These derivatives are captured by (3.68) and (3.71). According to (3.77), the welfare function is inverted U-shaped with respect to the output tax rate. The second-best output tax therefore reads

$$\tau_y^{sb} = \mathcal{E}'E' + c^* P' \frac{\partial \tilde{c}^*/\partial \tau_y}{\partial y^*/\partial \tau_y} \, . \tag{3.78}$$

According to Proposition 3.12, $\partial y^*/\partial \tau_y$ is negative. The sign of $\partial \tilde{c}^*/\partial \tau_y$ is indeterminate, in general. In contrast to second-best emissions taxation, it is not valid to conclude $\partial \tilde{c}^*/\partial \tau_y < 0$ in case of a linear cost function: Since second-best output taxation supposes a zero emissions tax ($\tau_\varepsilon = 0$), linear production costs ($K_{yy} = 0$) imply that condition C2 is satisfied. As already mentioned on p. 73, this condition is not compatible with an interior solution determined by (3.59)–(3.61). For that reason, we formulate the results with respect to second-best output taxation in a more general way.

Proposition 3.16. *In the long run, the second-best output tax satisfies*

$$\tau_y^{sb} \begin{Bmatrix} < \\ = \\ > \end{Bmatrix} \mathcal{E}'E' \qquad \Leftrightarrow \qquad \frac{\partial \tilde{c}^*}{\partial \tau_y} \begin{Bmatrix} < \\ = \\ > \end{Bmatrix} 0 \, .$$

The interpretation of this proposition is as follows. Setting the output tax equal to $\mathcal{E}'E'$ completely internalizes the marginal environmental damage. However, the stock of the durable good is still inefficiently low if the derivative $\partial \tilde{c}^*/\partial \tau_y$ is negative. Consequently, reducing the output tax below $\mathcal{E}'E'$ provides the firms with the incentive to increase output so that the stock raises and becomes more efficient. Hence, in this case, it is second-best optimal to set the output tax below $\mathcal{E}'E'$. The argumentation is reversed in case that the equilibrium stock increases in response to an output tax increase (i.e. $\partial \tilde{c}^*/\partial \tau_y > 0$). Nevertheless, a positive correlation between the durable stock and the output tax seems to be counterintuitive so that $\tau_y^{sb} < \mathcal{E}'E'$ is more plausible.

3.3 Practicability of Durable Goods Regulation

In this chapter, we have discussed the efficiency properties of different tax instruments in a polluting durable good economy. Our approach was mainly theoretical. We will now supplement the preceding analysis by discussing the practical relevance of the different tax instruments and the regulatory schemes derived.

Let us start with the tax levied on product durability. This tax rate constitutes the main difference to regulation in a nondurable good industry in which product durability is not a choice variable of producers. Unfortunately, the information requirements for employing the durability tax are extremely high. The tax basis of τ_ϕ is product durability and, hence, the regulator needs to know which factors influence product durability, e.g. how durable are different materials or which production process is suitable to render the products long living. Even if the regulator possesses all that information, incomplete information about the firms operations makes it difficult for her to find out which level of product durability manufactures actually choose. Perhaps she keeps track of the material composition of the products, but product durability may also be determined by the production technology employed, and producers typically have no incentive to reveal their production procedure. Furthermore, all regulatory schemes in Table 3.1 containing τ_ϕ require information about the decay function since they contain Ψ and Ψ'. This would not be a serious problem if the decay process is deterministic as in our abstract model analysis. However, in reality, decay is a highly stochastic process and finding out the relevant information to form reliable expectations about the lifetime of products is neither easy nor cheap. To sum up, from a practical point of view the tax rate τ_ϕ on product durability does not seem to be a suitable policy instrument for the regulator.

If we eliminate from Table 3.1 all those regulatory schemes containing a non-zero durability tax rate, then the schemes in rows 1–3, 5 and 6 remain. But it is not clear how to assess these remaining schemes from the viewpoint of practical policy. To see this, consider the stock tax. In principle, the stock of a durable good is easy to measure and to monitor. Moreover, some already established taxes may be interpreted as stock tax. An example is the German car tax ('Kraftfahrzeugsteuer') which is levied on car owners. Note, however, that the regulatory schemes in rows 5 and 6 of Table 3.1 do not contain a stock *tax*, but a stock *subsidy*. Due to structural impediments, such subsidies are often argued to be infeasible (e.g. Barnett, 1980; Simpson, 1995). Therefore, the regulator usually has to do regulation without both durability taxes and stock subsidies. The remaining instruments are emissions and output taxation.

Being constraint to the use of emissions and output taxation only, the regulator is still capable of restoring efficiency under perfect competition, indeed, but implementing the social welfare optimum under imperfect competition is no longer possible according to rows 5–9 of Table 3.1. Under imperfect competition, the regulator has to be content with the second-best welfare optimum which may be attained by setting either the emissions tax as in (3.76) or the output tax as in (3.78). If she sets the respective tax rate below the second-best level, then welfare under imperfect competition is still increased compared to the laissez faire situation; remember the inverted U-shape of the welfare function. An overinternalization result in second-best taxation, if it exists, would extend the scope for such welfare improvement since the maximum of the welfare function is attained at a higher tax rate.

It remains to clarify whether emissions or output taxes are more appropriate from a practical point of view. The tax basis of τ_ε are the physical emissions. During the last decades, engineers made great progress in developing technologies to monitor emissions. Nevertheless, it is still difficult to measure all emissions arising in the production process and, hence, applying emissions taxation may also be problematic. In contrast, the output of durable goods is easy to measure and to monitor since all market transactions are usually documented in the accounting system of firms. From a practical point of view, output taxation therefore seems to be the most appropriate means for the regulator to correct market failure in the polluting durable good economy.

4

Regulating Solid Waste of Consumer Durables

The previous section focused on emissions arising during the production process of the durable good. Another kind of pollution generated by durable consumption goods is solid consumption waste at the end of the products' life. This waste causes disposal costs which are of increasing importance in many countries. In an extensive survey for the Republic of Ireland, for example, Barrett and Lawlor (1997) report disposal costs of IR £ 11 to 14 per ton for waste collection, IR £ 94 per ton for (curbside) recycling, IR £ 45 per ton for waste incineration and IR £ 19 to 22 per ton for landfilling. Roughly speaking, the data for Ireland may serve as lower bounds for disposal costs in countries like the U.K. or Germany since owing to the low population density landfilling costs and external costs in Ireland are likely to be smaller than in other regions. Evidence for this assertion can be found in e.g. Brisson (1993), Miranda and Hale (1997) and Huhtala (1999).

The present chapter develops and investigates models incorporating solid consumption waste and the associated disposal costs. In Sect. 4.1, we first retain the assumption of commitment ability and investigate a vintage durable good model with consumption waste and producers who use open-loop strategies. Section 4.2 considers the case without precommitment and inquires into a two-period durable good model with consumption waste and producers who use closed-loop strategies. In both models, we first characterize the social welfare optimum (Sects. 4.1.1 and 4.2.1) and then determine the market equilibrium and the laissez faire allocation in order to make the case for governmental regulation (Sects. 4.1.2 and 4.2.2). Comparing the social optimum and the market equilibrium allows specifying regulatory schemes capable of restoring the efficient allocation (Sects. 4.1.3 and 4.2.3). For the case that the regulator does not dispose of all required instruments to correct for market failure, in Sects. 4.1.4 and 4.2.4, we conduct an analysis of second-best taxation.

In Sect. 4.3, we use the results of the theoretical models in order to evaluate practical waste policy. Policy instruments for solid waste management can basically divided into two groups. The first group consists of all instruments that intervene at the level of consumers or disposal firms. Examples are lump-sum or unit-based waste fees levied on households (Jenkins, 1993). The second group comprises all instruments which shift the burden of disposal costs from the public or consumers to producers. These instruments are consistent with the so-called principle of extended producers responsibility (EPR). Examples for such instruments are take-back requirements (Michaelis, 1995), public waste funds (Lidgren and Skogh, 1996) or privately organized waste management institutions like the German Green Dot Program (Rousso and Shah, 1994). On the basis of our theoretical analysis, the discussion of applied waste management issues in Sect. 4.3 will assess the impact of EPR-instruments and non-EPR instruments on product durability, waste generation and social welfare.

4.1 Durable Goods Markets with Commitment Ability of Producers

4.1.1 Model and Social Welfare Optimum

To analyze the case where producers are able to commit, we again choose a vintage model of a durable consumption good.[1] The model is given by

$$k_i(t) = K[y_i(t), \phi_i(t)] \qquad \forall\, i \in N \,, \tag{4.1}$$

$$c_i(t) = \int_{-\infty}^{t} D[t - v, \phi_i(v)] y_i(v) dv \qquad \forall\, i \in N \,, \tag{4.2}$$

$$p(t) = P[C(t)] \qquad \text{with} \quad C(t) = \sum_{j \in N} c_j(t) \,, \tag{4.3}$$

$$w_i(t) = -\int_{-\infty}^{t} D_a[t - v, \phi_i(v)] y_i(v) dv \qquad \forall\, i \in N \,, \tag{4.4}$$

$$e(t) = E[W(t)] \qquad \text{with} \quad W(t) = \sum_{j \in N} w_j(t) \,. \tag{4.5}$$

Equations (4.1)–(4.3) coincide with our basic durable good model presented in Sect. 2.2. At time t, firm i produces $y_i(t)$ units of the durable good with built-in product durability $\phi_i(t)$. In doing so, it incurs production costs represented by the production cost function (4.1) which is required to satisfy assumption A3. The output of firm i adds to the durable stock $c_i(t)$ given by (4.2). The decay function D is now supposed to satisfy not only assumptions A1, but also A2 with the consequence that in this section, we are allowed to apply the results of Lemma 2.1i and ii. The consumers' demand for the services of the

---

[1] The analysis in this section borrows heavily from Runkel (2003b).

durable stock is described by the demand function in (4.3). As in the previous chapters, we impose assumption A4 on this demand function.

In contrast to the models presented in Chap. 3, we now ignore emissions arising during the production of the durable good. Instead, the above model takes into account consumption waste emerging after the durable good is taken out of use. At time t, $-D_a[t - v, \phi_i(v)]y_i(v)$ units of firm i's vintage $v \leq t$ are scrapped and turn into consumption residuals. If the scrapped units of all vintages $v \leq t$ are added, we obtain $w_i(t)$ in (4.4) which represents firm i's amount of worn-out products at time t. As in the model of Sect. 3.2, it is supposed that an increase in product durability does not require additional material in the production process, but is attained, for example, by replacing less durable materials through more durable ones or by changing the product's design. As a consequence, all units of the durable good are of constant weight regardless of their durability and, thus, $w_i(t)$ in (4.4) stands not only for firm i's amount of scrapped units at time t, but also represents firm i's amount of solid waste at time t.

The industry amount of waste is given by $W(t)$ defined in (4.5). It equals the aggregate waste of all firms operating in the industry. Industry waste causes disposal costs which are measured by the twice continuously differentiable function $E : \mathbb{R}_+ \mapsto \mathbb{R}_+$ in (4.5). E encompasses all costs arising after the durable is taken out of use. It includes, for example, all private costs of waste collection, incineration, recycling and landfilling as well as all environmental costs entailed in these disposal activities (e.g. costs of smoke from incineration or costs of leakage). We assume $E'(W) > 0$ and $E''(W) \geq 0$ for all $W \in \mathbb{R}_+$ such that an increase in the industry waste raises disposal costs at non-decreasing rates.[2]

Two remarks are in order before proceeding with the analysis of the social optimum of the economy described by (4.1)–(4.5). First, to a large part, the subsequent analysis focuses on the stationary state again. As stationary state definition we will use

Definition D6. (y, ϕ, c, w) *is called asymptotic stationary state if*

$$\lim_{t \to \infty} y(t) = y , \quad \lim_{t \to \infty} \phi(t) = \phi , \quad \lim_{t \to \infty} c(t) = c \quad and \quad \lim_{t \to \infty} w(t) = w .$$

For simplicity, we again suppress the term 'asymptotic' throughout. Definition D6 is almost the same as definition D5 employed in Chaps. 2 and 3. The only difference is that we now additionally require the amount of solid waste to converge to a constant level w. In Sect. 2.2.2 on pp. 38, it was shown that the durable goods economy without solid waste approaches a stationary state in

---

[2] Note that $W(t)$ represents a waste *flow*. For simplicity, we do not explicitly account for waste accumulation and the associated waste *stock*. But the conjecture is that the main results of our analysis remain unchanged if a waste stock is taken into account. For the distinction of emissions stocks and flows, see e.g. Tietenberg (1996). A formal analysis of emissions stocks is contained in Xepapadeas (1997).

the sense of definition D5 if assumption A9 and condition (2.2) in assumption A1 are satisfied. The same conditions ensure that the durable goods waste economy considered in the present chapter attains a stationary state in the sense of definition D6. For notational convenience, we focus on a single firm and suppress the firm index i. The durable stock converges to $c = y\Gamma(\phi)$ for the same reasons given in Sect. 2.2.2 on p. 39. Under assumption A9, the amount of solid waste defined in (4.4) may be written as

$$\lim_{t \to \infty} w(t) = - \underbrace{\lim_{t \to \infty} \int_{-\infty}^{\hat{t}} D_a[t - v, \phi(v)]y(v)dv}_{=:\tilde{I}(t)}$$

$$- y \lim_{t \to \infty} \int_{\hat{t}}^{t} D_a(t - v, \phi)dv , \qquad (4.6)$$

where $\hat{t} \in \mathbb{R}_+$ is a point in time sufficiently large such that output and durability are already close to their constant levels y and ϕ, respectively. Condition (2.2) in assumption A1 implies $\tilde{I}(t) \to 0$ as $t \to \infty$. Consequently, if the transformation $a := t - v$ is employed, then (4.6) becomes

$$\lim_{t \to \infty} w(t) = -y \lim_{t \to \infty} \int_{0}^{t-\hat{t}} D_a(a, \phi)da = -y \int_{0}^{\infty} D_a(a, \phi)da = y =: w , \quad (4.7)$$

where we have made use of $\Lambda(\phi) = - \int_0^\infty D_a(a, \phi)da = 1$ according to Lemma 2.1. Hence, (y, ϕ, c, w) is a stationary state in the sense of definition D6. In this steady state, the number of newly produced units exactly equals the number of scrapped units, i.e. $w = y$. Rather than this one-to-one relation between output and waste, the subsequent analysis makes it necessary to consider

$$\tilde{w} := y\Omega(\phi) \qquad (4.8)$$

in some cases, for example, on p. 94. (4.8) represents the *approximate* long run amount of waste. To see this, observe that $\tilde{w}$ can be written as $\tilde{w} := w - y[1 - \Omega(\phi)]$. The approximation error $\tilde{\Delta} := y[1 - \Omega(\phi)]$ is positive so that the approximate long run amount of waste is smaller than the (exact) long run amount of waste. The approximation error is smaller, the smaller is the discount rate δ.

Second, before investigating the social optimum, it is necessary to clarify the impact of product durability on waste generation. Durability exerts two effects on consumption waste: For given output of vintage v, the cross derivative $-D_{a\phi}[a, \phi(v)]$ tells us how durability influences the waste quantity $-y(v)D_a[a, \phi(v)]$ of vintage v at age a. According to assumption A2, an increase in durability of vintage v prolongs the lifetime of some units of that vintage and implies less waste at relatively small age of the vintage $(-D_{a\phi}[a, \phi(v)] < 0$ for $a < A[\phi(v)])$ and more waste if the age exceeds a threshold value $(-D_{a\phi}[a, \phi(v)] > 0$ for $a > A[\phi(v)])$. However, Lemma 2.1i implies $\Lambda'(\phi) = \int_0^\infty D_{a\phi}[a, \phi(v)]da = 0$, i.e. over the whole lifetime of vintage v the sum over all changes in the amount of scrapped units has to be zero since every unit whose life is prolonged does not endure forever, but wears out in some future period. Hence, for any given output, increasing product durability has 'merely' a *waste-delaying effect*. It shifts waste generation to later periods, but leaves unaltered the total amount of waste caused by the vintage over its whole life. Besides this waste-delaying effect, raising product durability has a *waste-reducing effect* in the sense that it reduces the amount of waste necessary to provide a given stock of the durable good. To formalize this effect, focus on the stationary state and totally differentiate the long run stock $c = y\Gamma(\phi)$ as well as long run waste $w = y$. Setting $dc = 0$ yields

$$\left.\frac{dw}{d\phi}\right|_{dc=0} = -y\frac{\Gamma'(\phi)}{\Gamma(\phi)} < 0 . \tag{4.9}$$

(4.9) shows that, for given long run stock of the durable good, increasing long run product durability reduces the long run amount of solid waste. The reason is that output and durability are substitutes with respect to the durable stock. Hence, an increase in durability allows for reducing output so that waste declines. In the subsequent analysis, we will often express the waste-reducing effect in terms of the approximate amount of waste and the approximate stock of the durable good. Totally differentiating $\tilde{w} = y\Omega(\phi)$ from (4.8) and $\tilde{c} = y\Psi(\phi)$ from (3.39) and setting $d\tilde{c} = 0$ gives

$$\left.\frac{d\tilde{w}}{d\phi}\right|_{d\tilde{c}=0} = y\left[\Omega'(\phi) - \Omega(\phi)\frac{\Psi'(\phi)}{\Psi(\phi)}\right] < 0 . \tag{4.10}$$

The sign of this expression results from $\Omega' < 0$ and $\Omega, \Psi, \Psi' > 0$ proved in Lemma 2.1. Taking as given the approximate long run stock of the durable good, (4.10) states that an increase in long run product durability reduces the approximate long run amount of waste.

Let us now turn to characterizing the social welfare optimum. Imagine a social planner who maximizes the present value of social welfare. Instantaneous social welfare equals consumer benefits $S(C) := \int_0^C P(\xi)d\xi$ net of production and disposal costs. If the social planner focuses on a symmetric solution $x_i(t) = x(t)$ for all $x \in \{y, \phi, c, w\}$ and all $i \in N$, then she solves the problem

$$\left.\begin{aligned} \max_{\substack{y(0,\infty) \\ \phi(0,\infty)}} V &= \int_0^\infty \left[S[nc(t)] - nK[y(t), \phi(t)] - E[nw(t)]\right]e^{-\delta t}dt \\ \text{subject to} \quad c(t) &= \int_{-\infty}^t D[t - v, \phi(v)]y(v)dv , \\ \text{and} \quad w(t) &= -\int_{-\infty}^t D_a[t - v, \phi(v)]y(v)dv . \end{aligned}\right\} \tag{4.11}$$

According to (4.11), the social planner maximizes the present value of social welfare, V, taking into account the stock of the durable good and the amount of solid waste. (4.11) is an optimal control problem with integral state constraints. $y(t)$ and $\phi(t)$ are the control variables and $c(t)$ and $w(t)$ represent state variables. The present-value Hamiltonian reads

$$\mathcal{H}[y(t), \phi(t), c(t), w(t), \theta_c(t, \infty), \theta_w(t, \infty), t]$$

$$= \left[S[nc(t)] - nK[y(t), \phi(t)] - E[nw(t)] \right] e^{-\delta t}$$

$$+ \int_t^\infty D\big[v - t, \phi(t)\big] y(t) \theta_c(v) dv - \int_t^\infty D_a\big[v - t, \phi(t)\big] y(t) \theta_w(v) dv \ ,$$

where $\theta_c(t)$ and $\theta_w(t)$ are costate variables associated with the state variables $c(t)$ and $w(t)$. The costate variables may be interpreted as social shadow price of the durable stock and social shadow price of solid waste, respectively. Assuming an interior solution to (4.11) as well as a strictly concave Hamiltonian in the control variables and denoting socially optimal values by an 'o' yields

$$\theta_c(t) = \mathcal{H}_c(t) = nP[nc^o(t)]e^{-\delta t} \ , \tag{4.12}$$

$$\theta_w(t) = \mathcal{H}_w(t) = -nE'[nw^o(t)]e^{-\delta t} \ , \tag{4.13}$$

$$\mathcal{H}_y(t) = \mathcal{H}_\phi(t) = 0, \tag{4.14}$$

$$\mathcal{H}_{yy}(t) < 0 \ , \quad \mathcal{H}_{\phi\phi}(t) < 0 \ , \quad \mathcal{H}_{yy}(t)\mathcal{H}_{\phi\phi}(t) - \mathcal{H}_{y\phi}^2(t) > 0 \ , \tag{4.15}$$

for all $t \in [0, \infty[$. According to (4.12), the social shadow price of the durable stock is equal to marginal consumer benefits. Equation (4.13) states that the social shadow price of waste reflects marginal disposal costs and is negative since additional residuals are socially wasteful. Computing the derivatives in (4.14) and inserting (4.12) and (4.13) into the resulting expressions implies[3]

$$K_y(t) - \int_t^\infty D_a(v)E'(v)e^{-\delta(v-t)} dv = \int_t^\infty D(v)P(v)e^{-\delta(v-t)} dv \ , \tag{4.16}$$

$$\frac{K_\phi(t)}{y^o(t)} = \int_t^\infty D_\phi(v)P(v)e^{-\delta(v-t)} dv + \int_t^\infty D_{a\phi}(v)E'(v)e^{-\delta(v-t)} dv \ , \tag{4.17}$$

for all $t \in [0, \infty[$. The first term on the LHS of (4.16) equals marginal production costs of vintage t output. The second term on the LHS of (4.16) represents

---

[3] Note that the Jacobian determinant of the maximized Hamiltonian is negative definite since $\mathcal{H}_{cc}^{\max}(t) = n^2 P'(t)e^{-\delta t} < 0$, $\mathcal{H}_{ww}^{\max}(t) = -n^2 E''(t)e^{-\delta t} < 0$ and $\mathcal{H}_{cw}^{\max}(t) = 0$. Consequently, the maximized Hamiltonian is concave in the state variables and conditions (4.16) and (4.17) together with the integral constraints in (4.11) are necessary *and* sufficient for the welfare optimum.

marginal disposal costs of vintage t output. In contrast to marginal environmental costs of production emissions in the model of Sect. 3.2, marginal disposal costs of vintage t output arise not only at time t, but at times $v \geq t$
since a marginal increase in output of vintage t ceteris paribus leads to an increase in waste at t and some non-empty time interval immediately following
t. The term on the RHS of (4.16) stands for marginal consumer benefits of
vintage t output arising at $v \geq t$. In sum, equation (4.16) states that in the
first-best optimum, marginal social costs of vintage t output equal marginal
social benefits for all t. Analogously, (4.17) equates marginal social costs and
benefits of vintage t durability. The LHS of (4.17) gives marginal production costs of vintage t durability. The first term on the RHS of (4.17) equals
marginal consumer benefits of durability arising at $v \geq t$. The second term
on the RHS of (4.17) represents the social consequences of the waste-delaying
effect of product durability since it contains the cross-derivative $D_{a\phi}$: Shifting
waste to later periods is socially beneficial if marginal disposal costs E' are
non-increasing over time. In this case, the second term on the RHS of (4.17)
is unambiguously positive and, thus, represents marginal disposal benefits of
durability. To see this, remember assumption A2 and the strictly positive discount rate δ. However, it cannot be excluded that E' increases over time. The
second term on the RHS of (4.17) may therefore be negative such that the
waste-delaying effect of durability causes marginal disposal costs (!) instead
of benefits. This observation is of particular interest since in the applied literature it is usually argued that an increase in product durability reduces
disposal costs (e.g. Stahel, 1991). Our analysis demonstrates, however, that
social wastefulness of increasing durability cannot be excluded.

Considering the stationary state yields more clear-cut results. Using definition D6 and employing assumption A9 as well as condition (2.2) in assumption
A1, the efficiency conditions (4.16) and (4.17) become

$$K_y(y^o, \phi^o) + \Omega(\phi^o)E'(nw^o) = \Psi(\phi^o)P(nc^o) , \qquad (4.18)$$

$$\frac{K_\phi(y^o, \phi^o)}{y^o} = \Psi'(\phi^o)P(nc^o) - \Omega'(\phi^o)E'(nw^o) , \qquad (4.19)$$

where

$$c^o = y^o\Gamma(\phi^o) \qquad \text{and} \qquad w^o = y^o . \qquad (4.20)$$

(4.18) and (4.19) again equate marginal social costs and benefits of output
and durability, respectively. But now the waste-delaying effect of increasing
durability unambiguously generates marginal disposal benefits represented by
the term $-\Omega'E' > 0$ in (4.19). The reason is that E' is constant in the long
run and, thus, a unit of the durable good inflicts the same disposal costs
regardless of the time it is scrapped. Delaying waste by increasing durability
is then socially beneficial since society prefers postponing the generation of
waste owing to the positive discount rate.

Until now we have focused on the waste-delaying effect of product durability. To make visible the waste-reducing effect, solve (4.18) for P and insert P into (4.19). The result is

$$\frac{K_\phi(y^o, \phi^o)}{y^o} = \frac{\Psi'(\phi^o)}{\Psi(\phi^o)} K_y(y^o, \phi^o) - \left[\Omega'(\phi^o) - \Omega(\phi^o)\frac{\Psi'(\phi^o)}{\Psi(\phi^o)}\right] E'(nw^o). \quad (4.21)$$

(4.21) is an alternative way to express that in the social optimum, marginal social costs of durability equal marginal social benefits. As in (4.19), marginal social costs on the LHS of (4.21) consist of marginal production costs. But marginal social benefits are now based on the property that, for a given stock of the durable good, enhanced durability allows reducing output. This substitution has two effects on social costs. It saves costs of producing output. This effect is represented by the first term on the RHS of (4.21). Furthermore, the substitution of durability for output reduces disposal costs of waste since according to (4.10) the amount of waste declines. This effect is reflected by the second term on the RHS of (4.21) which is exactly the same as in (4.10).

To obtain more information on the social optimum, we conduct a comparative static analysis of the long run. Introducing a shift parameter γ with $E'_\gamma > 0$ and totally differentiating (4.18)–(4.20) yields

$$\frac{\partial y^o}{\partial \gamma} = \frac{E'_\gamma}{|J^o|}\left[ny^o\Gamma'P'\left(\Omega\Psi' - \Omega'\Psi\right) - \Omega\left(\frac{K_{\phi\phi}}{y^o} + \Omega''E' - \Psi''P\right)\right.$$

$$\left. + \Omega'\left(K_{y\phi} - \frac{K_\phi}{y^o}\right)\right], \quad (4.22)$$

$$\frac{\partial \phi^o}{\partial \gamma} = \frac{E'_\gamma}{|J^o|}\left[n\Gamma P'\left(\Omega'\Psi - \Omega\Psi'\right) - \Omega'K_{yy} + \frac{\Omega}{y^o}\left(K_{y\phi} - \frac{K_\phi}{y^o}\right)\right]. \quad (4.23)$$

The sign of the Jacobian determinant $|J^o|$ is indeterminate, in general. However, as in the previous chapters, we assume $|J^o| > 0$. (4.22) and (4.23) then imply

Proposition 4.1. *In the long run social optimum*

$$\frac{\partial y^o}{\partial \gamma} = \frac{\partial w^o}{\partial \gamma} < 0 \qquad and \qquad \frac{\partial \phi^o}{\partial \gamma} > 0. \quad (4.24)$$

Proof. In the long run, the concavity condition $\mathcal{H}_{\phi\phi}(t) < 0$ in (4.15) implies $K_{\phi\phi} + y^o\Omega''E' - y^o\Psi''P > 0$. From the Abel condition it follows $K_{y\phi} - K_\phi/y^o = K_{yy}K_\phi/K_y > 0$. The proof of Proposition 4.1 is then completed by (4.22), (4.23), Lemma 2.1, $w^o = y^o$, $|J^o| > 0$ and $E'_\gamma > 0$. $\qquad\square$

For increasing γ, every scrapped unit causes larger disposal costs. According to Proposition 4.1, the social planner reduces these additional disposal costs through substituting, at the margin, durability for output (waste-reducing effect of durability) and through delaying waste by an increase in durability (waste-delaying effect of durability). Therefore, efficient long run durability is the greater, the greater are marginal disposal costs. Note that we cannot expect this result also holds without qualifications along the adjustment path of the durable good economy. As shown in (4.17), outside the stationary state the waste-delaying effect of increasing durability may cause welfare costs rather than welfare benefits. If these costs overcompensate the welfare benefits of the waste-reducing effect, then the social planner might want to avoid additional disposal costs through reducing (!) durability.

4.1.2 Market Equilibrium and Laissez Faire

Consider now the case in which the durable is supplied by profit-maximizing firms. Profits equal revenues less costs and tax payments. Firms are assumed to pay taxes on each endogenous variable $x(t)$ with $x \in \{y, \phi, c, w\}$. The tax rates are $\tau_w(t)$ on solid waste, $\tau_c(t)$ on the stock of the durable good, $\tau_y(t)$ on output and $\tau_\phi(t)$ on product durability. These rates are not restricted in sign, and in case a tax rate is negative, it represents a subsidy instead of a tax. Firm i's pay-off is a function $\Pi^{ri} : \mathcal{S}^n \mapsto \mathbb{R}$ where

$$\Pi^{ri}(s_i, s_{-i}) = \int_0^\infty \left[P(\cdot) \int_{-\infty}^t D[t - v, \phi_i(v)] y_i(v) dv - K[y_i(t), \phi_i(t)] \right.$$

$$- \tau_y(t) y_i(t) - \tau_\phi(t) \phi_i(t) - \tau_c(t) \int_{-\infty}^t D[t - v, \phi_i(v)] y_i(v) dv$$

$$\left. + \tau_w(t) \int_{-\infty}^t D_a[t - v, \phi_i(v)] y_i(v) dv \right] e^{-\delta t} dt \qquad (4.25)$$

is the present value of firm i's rental profits over the planning period $[0, \infty[$. In deriving the rental equilibrium, we retain the assumption that producers possess commitment ability such that they face an open-loop information pattern. Rental and sales equilibria therefore coincide if either assumption A5 or assumption A6 is satisfied. As in the model of Sect. 3.2, this equivalence is not trivial since, in the rental case, stock and waste taxes are paid by producers whereas in the sales case, they are paid by consumers. Yet, rational utility-maximizing consumers subtract from their willingness-to-pay for the durable good their stock and waste tax payments.[4] Consequently, the stock and waste taxes enter the pay-off function of selling firms, too, and the open-loop sales equilibrium is the same as the open-loop rental equilibrium.

---

[4] This assertion is shown in Chap. 5 where we consider a representative utility-maximizing consumer. See especially footnote 7 on p. 153.

For characterizing the open-loop equilibrium as defined in definition D2, we have to consider the firms' profit maximization problems. Firm i chooses time paths of output and durability in order to maximize rental profits taking as given the time paths of its rivals' strategies. This maximization problem can be written as

$$\left.\begin{array}{l} \max_{\substack{y_i(0,\infty) \\ \phi_i(0,\infty)}} \int_0^\infty \left[c_i(t)P[c_i(t) + C^i(t)] - K[y_i(t), \phi_i(t)] - \tau_y(t)y_i(t) \right. \\ \\ \qquad\qquad \left. - \tau_\phi(t)\phi_i(t) - \tau_c(t)c_i(t) - \tau_w(t)w_i(t) \right] e^{-\delta t} dt \\ \\ \text{subject to} \quad c_i(t) = \int_{-\infty}^t D[t - v, \phi_i(v)]y_i(v)dv , \\ \\ \text{and} \quad w_i(t) = -\int_{-\infty}^t D_a[t - v, \phi_i(v)]y_i(v)dv . \end{array}\right\} \quad (4.26)$$

The present-value Hamiltonian of this optimal control problem with integrals state constraints is

$$\mathcal{H}[y_i(t), \phi_i(t), c_i(t), w_i(t), \lambda_{ci}(t, \infty), \lambda_{wi}(t, \infty), t]$$

$$= \left[c_i(t)P[C(t)] - K[y_i(t), \phi_i(t)] - \tau_y(t)y_i(t) \right.$$

$$\left. - \tau_\phi(t)\phi_i(t) - \tau_c(t)c_i(t) - \tau_w(t)w_i(t) \right] e^{-\delta t}$$

$$+ \int_t^\infty D[v - t, \phi_i(t)]y_i(t)\lambda_{ci}(v)dv$$

$$- \int_t^\infty D_a[v - t, \phi_i(t)]y_i(t)\lambda_{wi}(v)dv ,$$

where $\lambda_{ci}(t)$ and $\lambda_{wi}(t)$ are costate variables associated with the state variables $c_i(t)$ and $w_i(t)$, respectively. We assume that the Hamiltonian is strictly concave in the control variables $[y_i(t), \phi_i(t)]$ and focus on an interior solution. The optimality conditions for the solution of (4.26) may be rearranged to yield

$$K_y(t) + \tau_y(t) - \int_t^\infty D_a(v)\tau_w(v)e^{-\delta(v-t)}dv + \int_t^\infty D(v)\tau_c(v)e^{-\delta(v-t)}dv$$

$$= \int_t^\infty D(v)[P(v) + c_i^*(v)P'(v)]e^{-\delta(v-t)}dv , \quad (4.27)$$

$$\frac{K_\phi(t) + \tau_\phi(t)}{y_i^*(t)} + \int_t^\infty D_\phi(v)\tau_c(v)e^{-\delta(v-t)}dv$$

$$= \int_t^\infty D_\phi(v)[P(v) + c_i^*(v)P'(v)]e^{-\delta(v-t)}\,dv$$

$$+ \int_t^\infty D_{a\phi}(v)\tau_w(v)e^{-\delta(v-t)}\,dv \tag{4.28}$$

for all $t \in [0, \infty[$. A star indicates profit-maximizing values of the respective variable. According to (4.27) and (4.28), firm i sets output and durability such that marginal costs equal marginal revenues. The terms on the LHS of (4.27) equal marginal costs of vintage t output (marginal production costs plus marginal output, waste and stock tax payments) and the RHS represents marginal revenues of vintage t output. The terms on the LHS of (4.28) are the firm's marginal costs of durability (marginal production costs plus marginal durability and stock tax payments) whereas the RHS captures marginal revenues and waste tax payments of durability. Note that the second term on the RHS of (4.28) shows how the waste-delaying effect of enhanced durability influences waste tax payments. Analogous to our argumentation in the analysis of the social optimum, delaying waste reduces firm i's waste tax payments if the waste tax rate τ_w is non-increasing over time. However, if the waste tax rate does increase in time, then delaying waste may raise firm i's waste tax payments.

More clear-cut results are obtained in the stationary state defined in definition D6. Focusing on a symmetric long run market equilibrium, (4.27) and (4.28) simplify to

$$K_y(y^*, \phi^*) + \tau_y + \tau_w\Omega(\phi^*) + \tau_c\Psi(\phi^*) = \Big[P(nc^*) + cP'(nc^*)\Big]\Psi(\phi^*), \tag{4.29}$$

$$\frac{K_\phi(y^*, \phi^*) + \tau_\phi}{y^*} + \tau_c\Psi'(\phi^*) = \Big[P(nc^*) + cP'(nc^*)\Big]\Psi'(\phi^*) - \tau_w\Omega'(\phi^*), \tag{4.30}$$

with

$$c^* = y^*\Gamma(\phi^*) \qquad \text{and} \qquad w^* = y^*, \tag{4.31}$$

where the star indicates the long run market equilibrium allocation. As the general conditions (4.27) and (4.28), also (4.29) and (4.30) equate marginal costs and revenues of output and durability. However, τ_w is constant in the long run and increasing durability is profitable for the firm since the waste-delaying effect reduces the present value of waste tax payments due to the positive discount rate, i.e. $-\tau_w\Omega' > 0$.

Not only the waste-delaying effect, but also the waste-reducing effect of increased durability curtails the firm's waste tax payments in the long run. To show this, we solve (4.29) for $P + cP'$ and insert this term into (4.30). The result is

$$\frac{K_\phi(y^*, \phi^*) + \tau_\phi}{y^*} = \frac{\Psi'(\phi^*)}{\Psi(\phi^*)}K_y(y^*, \phi^*) + \frac{\Psi'(\phi^*)}{\Psi(\phi^*)}\tau_y$$

$$- \left[\Omega'(\phi^*) - \Omega(\phi^*) \frac{\Psi'(\phi^*)}{\Psi(\phi^*)} \right] \tau_w . \quad (4.32)$$

Equation (4.32) is an alternative condition for equilibrium durability. Like (4.30) it balances the firms' advantages (RHS) and disadvantages (LHS) of a marginal increase in durability. In contrast to (4.30), however, the advantages now do not reflect marginal revenues plus saved waste tax payments from delaying waste, but comprise saved production costs and saved output tax payments from reducing output (first two terms on the RHS) as well as saved waste tax payments from reducing waste (third term on the RHS). Note that the bracketed term in (4.32) equals the waste-reducing effect of enhanced durability in (4.10).

A comparative static analysis of the steady state answers the question of whether Swan's independence result is valid. Totally differentiating (4.29)–(4.31) and employing Cramer's rule yields

$$\frac{\partial \phi^*}{\partial n} = \frac{c^*(P' + c^* P'')}{|J^*|} \left[\frac{K_\phi + \tau_\phi + y^* \tau_w \Omega'}{P + c^* P' - \tau_c} K_{yy} \right.$$

$$\left. - \frac{K_y + \tau_y + \tau_w \Omega}{P + c^* P' - \tau_c} \frac{K_{yy} K_\phi}{K_y} + \frac{\tau_\phi \Psi}{y^*} \right]. \quad (4.33)$$

The sign of the Jacobian determinant $|J^*|$ is ambiguous, in general. However, it can be shown that a positive Jacobian determinant is a necessary and sufficient condition for the existence of a second-best tax.[5] We therefore assume $|J^*| > 0$ in the subsequent analysis. In Proposition 3.11 of Sect. 3.2, it was proved that the independence result is upset by emissions, output and durability taxation while stock taxation does not influence product durability. Replacing emissions taxation by waste taxation, (4.33) shows that this result carries over to our durable good economy with consumption waste.

If we set $\tau_c = \tau_\phi = 0$, then the comparative static analysis also yields

$$\frac{\partial y^*}{\partial \tau_w} = \frac{1}{|J^*|} \left[y^* \Gamma' \mathcal{X} \left(\Omega \Psi' - \Omega' \Psi \right) + \Omega' \left(K_{y\phi} - \frac{K_\phi}{y^*} \right) \right.$$

$$\left. - \Omega \left(\frac{K_{\phi\phi}}{y^*} + \tau_w \Omega'' - \Psi''(P + c^* P') \right) \right], \quad (4.34)$$

$$\frac{\partial \phi^*}{\partial \tau_w} = \frac{1}{|J^*|} \left[\Gamma \mathcal{X} \left(\Omega' \Psi - \Omega \Psi' \right) - \Omega' K_{yy} + \frac{\Omega}{y^*} \left(K_{y\phi} - \frac{K_\phi}{y^*} \right) \right], \quad (4.35)$$

and

$$\frac{\partial y^*}{\partial \tau_y} = -\frac{1}{|J^*|}\left[\frac{K_{\phi\phi}}{y^*} + \tau_w \Omega'' - \Psi''(P + c^*P') - y^*\Psi'\Gamma'\mathcal{X}\right], \quad (4.36)$$

$$\frac{\partial \phi^*}{\partial \tau_y} = -\frac{1}{|J^*|}\left[\Gamma\Psi'\mathcal{X} - \left(K_{y\phi} - \frac{K_\phi}{y^*}\right)\right], \quad (4.37)$$

where $\mathcal{X} = (n+1)P' + nc^*P'' < 0$ under imperfect competition and $\mathcal{X} = nP' < 0$ under perfect competition. (4.34)–(4.37) imply

Proposition 4.2. *Set $\tau_c = \tau_\phi = 0$ and let $\tau \in \{\tau_y, \tau_w\}$. Then, in the long run market equilibrium*

$$\frac{\partial y^*}{\partial \tau} = \frac{\partial w^*}{\partial \tau} < 0, \qquad \frac{\partial \tilde{w}^*}{\partial \tau} < 0, \qquad \frac{\partial \phi^*}{\partial \tau} > 0, \qquad \frac{\partial \tilde{c}^*}{\partial \tau} \gtreqless 0.$$

This is true under both perfect and imperfect competition.

Proof. In the long run, the concavity condition $\mathcal{H}_{\phi\phi}(t) < 0$ reads $K_{\phi\phi} + y^*\tau_w\Omega'' - y^*\Psi''(P+c^*P') > 0$ and the Abel condition implies $K_{y\phi} - K_\phi/y^* = K_{yy}K_\phi/K_y > 0$. Taking into account Lemma 2.1, $\mathcal{X} > 0$ and (4.34)–(4.37) proves $\partial y^*/\partial \tau < 0$ and $\partial \phi^*/\partial \tau > 0$. $\partial w^*/\partial \tau < 0$ follows from $w^* = y^*$. From (4.8) we obtain

$$\frac{\partial \tilde{w}^*}{\partial \tau} = \Omega\frac{\partial y^*}{\partial \tau} + y^*\Omega'\frac{\partial \phi^*}{\partial \tau} < 0.$$

Finally, $\partial \tilde{c}^*/\partial \tau \gtreqless 0$ follows from $\partial \tilde{c}^*/\partial \tau = \Psi(\partial y^*/\partial \tau) + y^*\Psi'(\partial \phi^*/\partial \tau)$. □

Waste taxation increases the firms' disposal costs. In response to such a tax, firms substitute durability for output in order to curtail and delay waste (both w^* and $\tilde{w}^*$) and thus to reduce the present value of waste tax payments. Since changes in output and durability point to opposite directions, the change in the (approximate) stock of the durable good is indeterminate, in general. The effects of output taxation on long run equilibrium values are the same as those of waste taxation, but they possess a different intuition. An output tax provides firms with the incentive to reduce output. Since output and durability are substitutes with respect to the durable's stock, the firms compensate, to some extent, the reduction in output by enhancing durability. The changes in durability and output induce waste to decrease whereas the effect on the stock of the durable is ambiguous.

Before turning to the study of efficient regulation in the durable good industry with consumption waste, we wish to make the case for regulation by identifying market failure under laissez faire. For simplicity, our focus is on the stationary state only. It then follows

Proposition 4.3. *Set $\tau_w = \tau_y = \tau_c = \tau_\phi = 0$. Then,*
(i) the long run market equilibrium under perfect competition satisfies

$$y^* = w^* > y^o = w^o \,, \qquad \phi^* < \phi^o \,, \qquad \tilde{w}^* > \tilde{w}^o \quad and \quad \tilde{c}^* \gtreqqless \tilde{c}^o \,,$$

(ii) the long run market equilibrium under imperfect competition satisfies

$$y^* = w^* \gtreqqless y^o = w^o \,, \qquad \phi^* < \phi^o \,, \qquad \tilde{w}^* \gtreqqless \tilde{w}^o \quad and \quad \tilde{c}^* \gtreqqless \tilde{c}^o \,,$$

where ϕ^* is the same as in the long run market equilibrium under perfect competition.

Proof. Table 4.1 on p. 102 shows that $y^* = w^* = y^o = w^o$, $\phi^* = \phi^o$, $\tilde{w}^* = \tilde{w}^o$ and $\tilde{c}^* = \tilde{c}^o$ if $\tau_w = E' > 0$ and $\tau_y = \tau_c = \tau_\phi = 0$. This information together with Proposition 4.2 proves Proposition 4.3i. Swan's independence result implies $\phi^* < \phi^o$ in Proposition 4.3ii. According to Table 4.1, under imperfect competition, $y^* = w^* = y^o = w^o$, $\tilde{w}^* = \tilde{w}^o$ and $\tilde{c}^* = \tilde{c}^o$ if $\tau_w = E' > 0$, $\tau_c = c^* P' < 0$ and $\tau_y = \tau_\phi = 0$. Since $\partial y^* / \partial \tau_c$ is indeterminate in sign, the remainder of Proposition 4.3ii follows. $\qquad\qquad\Box$

The intuition of this result is as follows. Let us start with the case of perfect competition considered in Proposition 4.3i. An increase in long run product durability has the social benefit of reducing disposal costs by delaying and reducing waste. Under laissez faire, producers ignore this benefit since disposal costs are external to them and, consequently, they choose product durability inefficiently low. Analogously, choosing long run output at a higher level causes additional social costs since waste and disposal costs increase. Unregulated producers do not account for this effect and therefore choose output inefficiently large. The distorted choice of output and durability renders waste inefficiently large (both w^* and $\tilde{w}^*$) whereas the stock of the durable may be smaller or larger than its first-best level.

To give the intuition of Proposition 4.3ii, note that under imperfect competition there is a further market imperfection, namely the market power of producers. Market power provides producers with the incentive to choose long run output below the efficient level. This effect works in opposite direction as the impact of external disposal costs. Hence, output and waste are not necessarily too large as in case of perfect competition, but may also be inefficiently low. In contrast, long run product durability continues to fall short of its first-best level. The reason is that durability is independent of market structure, i.e. Swan's independence result which holds in the present model due to the Abel condition.

Proposition 4.3 exclusively focuses on the stationary state. On the adjustment path, it cannot be ruled out that the underprovision of durability identified in Proposition 4.3 turns into an overprovision of durability. As argued in the context of interpreting (4.17), enhancing durability may *increase* disposal costs outside the steady state so that increasing durability causes a negative externality and unregulated producers choose durability inefficiently high. Unfortunately, we are not able to formally prove this assertion owing to the complexity of adjustment dynamics.

4.1.3 First-Best Regulation

The next step is to investigate how the market failure identified in Proposition 4.3 may be corrected. Comparing the efficiency conditions (4.16) and (4.17) with the market equilibrium conditions (4.27) and (4.28) reveals that under perfect competition ($P' \equiv 0$) efficiency is attained by setting $\tau_w(t) = E'(t) > 0$ and $\tau_y(t) = \tau_c(t) = \tau_\phi(t) = 0$ for all $t \in [0, \infty[$. Hence, in the durable good economy with consumption waste, too, Pigouvian taxation is appropriate to implement the efficient allocation under perfect competition. The Pigouvian waste tax induces producers to account for the impact of durability and output on solid waste. Separate instruments to correct both distorted durability and output are not required since both inefficiencies lead to an inappropriate amount of waste and, therefore, may be corrected simultaneously by one and the same instrument.[6] Under imperfect competition ($P' < 0$), the comparison of the efficiency conditions (4.16) and (4.17) with the equilibrium conditions (4.27) and (4.28) shows that efficiency is restored by setting $\tau_w(t) = E'(t) > 0$, $\tau_c(t) = c^o(t)P'(t) < 0$ and $\tau_y(t) = \tau_\phi(t) = 0$ for all $t \in [0, \infty[$. The intuition of this tax-subsidy scheme is likewise straightforward. The scheme contains a Pigouvian waste tax that eliminates the disposal costs externality and a stock subsidy which corrects the distortion due to market power of firms.

Besides these two straightforward policies there is a variety of further regulatory schemes which restore efficiency. Focusing on the long run and comparing (4.16) and (4.17) with (4.27) and (4.28) yields

Proposition 4.4. *The long run market equilibrium is socially optimal if and only if*

$$\tau_y + \left[\tau_w - E'(nw^o)\right]\Omega(\phi^o) = \left[c^o P'(nc^o) - \tau_c\right]\Psi(\phi^o) \,,$$

$$\frac{\tau_\phi}{y^o} + \left[\tau_w - E'(nw^o)\right]\Omega'(\phi^o) = \left[c^o P'(nc^o) - \tau_c\right]\Psi'(\phi^o) \,.$$

Not surprisingly, there is a continuum of possibilities to set the tax rates appropriately since the number of available instruments exceeds the number of equations to be satisfied. From a viewpoint of public choice, this is a somewhat unsatisfactory result. It is more interesting to consider some special cases of Proposition 4.4. Table 4.1 therefore contains all those regulatory schemes which comprise at most two non-zero tax rates. Let us start with the case of perfect competition. Rows 1–4 show that Pigouvian taxation is the only efficient policy scheme containing one and only one non-zero tax rate. This finding is in contrast to the results of Table 3.1 which showed that in the durable good economy with production emissions, output taxation is an

---

[6] In this sense, there is a deviation from the Tinbergen-principle which states that for every distortion the regulator needs exactly one instrument to attain efficiency. See Tinbergen (1967).

Table 4.1. First-Best Regulatory Schemes in the Presence of Consumption Waste

row	τ_w	τ_y	τ_c	τ_ϕ
			Perfect competition	
1	$E' > 0$	–	–	–
2	–	$\Omega E' > 0$	–	$y^o\Omega'E' < 0$
3	–	–	$\Omega E'/\Psi > 0$	$(\Omega' - \Omega\Psi'/\Psi)E' < 0$
4	–	$(\Omega - \Omega'\Psi/\Psi')E' > 0$	$\Omega'E'/\Psi' < 0$	–
			Imperfect competition	
5	$E' > 0$	–	$c^o P' < 0$	–
6	$E' + c^o\Psi'P'/\Omega' \gtreqless 0$	$c^o(\Psi - \Psi'\Omega/\Omega')P' < 0$	–	–
7	$E' + c^o\Psi P'/\Omega \gtreqless 0$	–	–	$y^o c^o(\Psi' - \Psi\Omega'/\Omega)P' < 0$
8	–	$(\Omega - \Omega'\Psi/\Psi')E' > 0$	$\Omega'E'/\Psi' + c^o P' < 0$	–
9	–	$\Omega E' + c^o\Psi P' \gtreqless 0$	–	$y^o\Omega'E' + c^o\Psi'P' < 0$
10	–	–	$c^o P' + \Omega E'/\Psi \gtreqless 0$	$y^o(\Omega' - \Omega\Psi'/\Psi)E' < 0$

additional appropriate means to render socially optimal the market equilibrium under perfect competition. The reason for this difference is that in the durable good economy with production emissions, the amount of pollutants measured by the emissions function (3.42) depends on output only whereas in the durable good economy with consumption waste, the scrapped durables defined in (4.4) are influenced by both output and product durability. This is why in the presence of consumption waste an output tax alone is not sufficient to achieve the welfare optimum.

Row 2 of Table 4.1 shows that under perfect competition efficiency may be restored if an output tax is combined with a durability subsidy. The intuition of this tax-subsidy scheme is as follows. The output tax reflects the present value of disposal costs incurred by each unit of the durable good over its whole lifetime ($\Omega E'$). This tax brings the inefficiently large output down to its efficient level. To implement the efficient amount of waste, the regulator then still needs to raise durability. In doing so, she ought to leave output unchanged to avoid that it will become inefficient again. For constant output, increasing durability has a waste-delaying effect only and, thus, τ_ϕ in row 2 of Table 4.1 reflects marginal disposal benefits of this waste-delaying effect ($y^o\Omega'E'$), but not marginal disposal benefits of the waste-reducing effect. Outside the stationary state, it is not clear whether the optimal τ_ϕ is a tax or a subsidy since disposal benefits of the waste-delaying effect may be of either sign. In the long run, however, delaying waste unambiguously reduces the present value of disposal costs and τ_ϕ in row 2 of Table 4.1 is negative. Consequently, long run durability has to be subsidized so that producers receive the incentive for increasing durability to its efficient level.

Another possibility to restore efficiency under perfect competition is the regulatory scheme defined in row 3 of Table 4.1. This scheme replaces the output tax in row 2 of Table 4.1 by a stock tax rendering the durable stock efficient. The regulator then implements the efficient values of the other variables if she shifts durability to its efficient level. But in doing so, she has to leave the stock constant since otherwise it would become inefficient again. For a constant stock, enhancing durability has a waste-reducing effect only and, thus, the tax rate on durability now contains the associated marginal disposal benefits rather than the marginal disposal benefits of the waste-delaying effect. To see this, note that the bracketed term in τ_ϕ equals the waste-reducing effect in (4.10). Since marginal disposal benefits of the waste-reducing effect are positive, τ_ϕ is a durability subsidy that induces firms to increase durability to its efficient level.

A further policy scheme capable of implementing the social optimum under perfect competition is the combination of an output tax and a stock subsidy defined in row 4 of Table 4.1. The output tax accounts for marginal disposal costs E' and induces durable good producers to reduce both output and solid waste. As argued above, however, an output tax alone is not sufficient to correct the disposal externality since it leads to an underprovision of the durable stock. This underprovision has to be removed by a subsidy on the stock of the durable good.

Rows 5–10 of Table 4.1 contain efficiency restoring regulatory schemes for the case of imperfect competition. The extensive discussion of tax-subsidy schemes in the upper part of Table 4.1 can be applied here with some appropriate modifications. We therefore omit the description of efficiency-restoring policies in the case of imperfect competition except for drawing attention to two general properties. First, under imperfect competition, it is not possible to attain efficiency with one non-zero tax rate only. As in the production emissions model of Sect. 3.2, the reason for this result is that the regulator has to account not only for the disposal externality, but also for the market imperfection due to market power of durable good producers. Second, all regulatory schemes listed in rows 5–10 of Table 4.1 contain at least one subsidy. As discussed in Sect. 3.3, subsidies to imperfectly competitive firms are problematic from a practical point of view since they are often argued to be politically infeasible. In such cases, the regulator is not able to achieve the efficient allocation under imperfect competition and has to settle for the second-best optimum.

4.1.4 Second-Best Taxation

In deriving the second-best optimum, we consider the cases in which the regulator has at her disposal either waste regulation or output regulation, but not both. Furthermore, the focus is on the stationary state. As shown in Sect. 2.2.2, using the state constraint $c = y\Psi(\phi) - \nu_1\Psi(\nu_2) + \nu_1\Gamma(\nu_2)$ with given ν_1 and ν_2 rather than the state constraint $c = y\Gamma(\phi)$ ensures that the long run maximization takes into account the adjustment path of the durable good economy. It is straightforward to show that for the durable good economy with consumption waste, we also need to replace the constraint $w = y$ by the constraint $w = y\Omega(\phi) - \nu_3\Omega(\nu_4) + \nu_3$. The parameter ν_3 and ν_4 are taken as given in differentiating the objective function. In the optimum conditions, they are set equal to the long run equilibrium values of output y^* and durability ϕ^*, respectively.

Investigating second-best waste taxation requires setting $\tau_y = \tau_c = \tau_\phi = 0$. The regulator then solves the problem

$$\left. \begin{aligned} \max_{\tau_w} V(\tau_w) &= S[nc^*(\tau_w)] - nK[y^*(\tau_w), \phi^*(\tau_w)] - E[nw^*(\tau_w)] \\ \text{subject to} \quad c^*(\tau_w) &= y^*(\tau_w)\Psi[\phi^*(\tau_w)] - \nu_1\Psi(\nu_2) + \nu_1\Gamma(\nu_2) \\ \text{and} \quad w^*(\tau_w) &= y^*(\tau_w)\Omega[\phi^*(\tau_w)] - \nu_3\Omega(\nu_4) + \nu_3 . \end{aligned} \right\} \quad (4.38)$$

Hence, the regulator maximizes social welfare in the long run market equilibrium and takes into account constraints for the long run stock of the durable good and long run waste. Using the long run equilibrium conditions (4.29)–(4.31) as well as the definitions of the approximate stock of the durable good in (3.39) and the approximate amount of consumption waste in (4.8) yields

$$V'(\tau_w) \gtreqqless 0 \qquad \Leftrightarrow \qquad \tau_w \lesseqqgtr E' + c^* P' \frac{\partial \tilde{c}^*/\partial \tau_w}{\partial \tilde{w}^*/\partial \tau_w} \ . \qquad (4.39)$$

According to (4.39), long run social welfare increases for all τ_w smaller than the threshold value $E' + c^* P'(\partial \tilde{c}^*/\partial \tau_w)/(\partial \tilde{w}^*/\partial \tau_w)$ and decreases for all τ_w greater than the threshold value. Hence, the second-best waste tax reads[7]

$$\tau_w^{sb} = E' + c^* P' \frac{\partial \tilde{c}^*/\partial \tau_w}{\partial \tilde{w}^*/\partial \tau_w} \ .$$

Under perfect competition ($P' \equiv 0$), the second-best waste tax coincides with the first-best one determined in row 1 of Table 4.1. But under imperfect competition, we have $P' < 0$ and the second term in τ_w^{sb} is different from zero, in general. According to Proposition 4.2, solid waste declines with an increasing waste tax rate ($\partial \tilde{w}^*/\partial \tau_w < 0$) whereas the impact of the waste tax rate on the durable stock is ambiguous ($\partial \tilde{c}^*/\partial \tau_w \gtreqqless 0$). We therefore obtain

Proposition 4.5. *Under imperfect competition, the second best waste tax satisfies*

$$\tau_w^{sb} \begin{Bmatrix} > \\ = \\ < \end{Bmatrix} E' \qquad \Leftrightarrow \qquad \frac{\partial \tilde{c}^*}{\partial \tau_w} \begin{Bmatrix} > \\ = \\ < \end{Bmatrix} 0 \ .$$

If the stock of the durable good increases as the waste tax increases, overinternalization is second-best optimal in the durable goods economy with consumption waste. As argued already in Chap. 3, however, a negative correlation between the equilibrium stock of the durable good and the waste tax rate seems to be plausible and empirical relevant. In this case, the second-best waste tax falls short of marginal disposal costs. This underinternalization result has a good economic intuition: Suppose solid waste is taxed by a rate equal to marginal disposal costs. This completely internalizes external disposal costs, indeed, but output is still smaller than in the first-best optimum since producers take advantage of their market power and charge an inefficiently high price. Reducing the waste tax below marginal disposal costs entails the welfare gain of moving output closer to the efficient level and underinternalization becomes second-best optimal.

Finally, consider second-best output taxation. This issue is interesting not only under imperfect competition, but also under perfect competition since the rows 1–4 of Table 4.1 show that output taxation alone is not sufficient to attain the efficient allocation under perfect competition. Analogous to equation (4.39), setting $\tau_w = \tau_c = \tau_\phi = 0$ yields

$$V'(\tau_y) \gtreqqless 0 \quad \Leftrightarrow \quad \tau_y \lesseqqgtr \Omega E' + y^* \Omega' E' \frac{\partial \phi^*/\partial \tau_y}{\partial y^*/\partial \tau_y} + c^* P' \frac{\partial \tilde{c}^*/\partial \tau_y}{\partial y^*/\partial \tau_y} \ . \quad (4.40)$$

---

[7] As in footnote 12 on p. 82, equation (4.39) can be used to show that $|J^*| > 0$ is necessary and sufficient for the existence of a second-best tax.

Hence, the second-best output tax reads

$$\tau_y^{sb} = \Omega E' + y^* \Omega' E' \frac{\partial \phi^* / \partial \tau_y}{\partial y^* / \partial \tau_y} + c^* P' \frac{\partial \tilde{c}^* / \partial \tau_y}{\partial y^* / \partial \tau_y} \; .$$

When combined with Proposition 4.2, this result immediately leads to

Proposition 4.6. *Under perfect competition, the second-best output tax satisfies $\tau_y^{sb} > \Omega E'$. Under imperfect competition, the second-best output tax satisfies*

$$\tau_y^{sb} \left\{ \begin{matrix} > \\ \geqq \\ < \end{matrix} \right\} E' \Omega \qquad if \qquad \frac{\partial \tilde{c}^*}{\partial \tau_y} \left\{ \begin{matrix} \geq \\ \\ < \end{matrix} \right\} 0 \; .$$

Proposition 4.6 states that the second-best output tax under perfect competition exceeds the term $\Omega E'$ which represents the present value of marginal disposal costs caused by a unit of the durable good over its whole lifetime. The intuition is as follows: If output is taxed by its marginal damage $\Omega E'$, then output would be efficient, as row 2 of Table 4.1 shows, *provided* durability is also at its efficient level. But durability tends to be inefficiently low since it is not subsidized by its marginal disposal benefits. Raising the output tax rate beyond $\Omega E'$ mitigates this distortion because durability increases and gets closer to its efficient level according to Proposition 4.2. Hence, the second-best output tax under perfect competition exceeds $\Omega E'$. Under imperfect competition, the determinants of the second-best output tax are more complex. The tax rate may be greater or smaller than the present value of marginal disposal costs since it comprises an additional term which captures the firms' market power. In case that the equilibrium stock of the durable good is non-decreasing in the output tax, the second-best output tax unambiguously exceeds the present value $\Omega E'$ of disposal costs caused by a unit of the durable good.

4.2 Durable Goods Markets without Commitment Ability of Producers

4.2.1 Model and Social Welfare Optimum

In all models discussed above, it has been assumed that durable good producers possess commitment ability. They use open-loop strategies and choose the whole time path of their decision variables right at the beginning of the planning period. This section drops the assumption of commitment ability, i.e. we now assume that durable good producers use closed-loop strategies and recalculate their initial plans in every period. We will keep considering an economy with consumption waste of consumer durables although the major insights of the analysis can be shown to be valid in a durable good economy with production emissions, too.[8]

From Bulow (1986) and Goering (1992), it is well known that the vintage approach presented in the previous sections is not tractable for analyzing durable good producers without commitment ability. We therefore resort to and modify in an appropriate way the two-period model put forward by these authors in order to inquire into environmental policy in the absence of commitment ability. The model is given by

$$k_{1i} = \kappa(\phi_i)y_{1i}, \qquad k_{2i} = \bar{\kappa}y_{2i} \qquad \forall\, i \in N\,, \tag{4.41}$$

$$c_{1i} = y_{1i}, \qquad c_{2i} = y_{2i} + \phi_i y_{1i} \qquad \forall\, i \in N\,, \tag{4.42}$$

$$p_t = P(C_t) \quad \text{with} \quad C_t = \sum_{j \in N} c_{tj} \qquad \forall\, t \in \{1,2\}\,, \tag{4.43}$$

$$w_{1i} = (1 - \phi_i)c_{1i}, \qquad w_{2i} = c_{2i} \qquad \forall\, i \in N\,, \tag{4.44}$$

$$e_t = E(W_t) \quad \text{with} \quad W_t = \sum_{j \in N} w_{tj} \qquad \forall\, t \in \{1,2\}\,. \tag{4.45}$$

There are again $n \geq 1$ firms operating in the durable good industry. In period $t \in \{1,2\}$, firm $i \in N$ produces y_{ti} units of the durable good. Product durability of period 1 units is $\phi_i \in [0,1]$. All firms face the same production costs described by (4.41). For simplicity, the cost function is assumed to be linear in output. In the first period, unit costs are a function $\kappa : \mathbb{R}_+ \mapsto \mathbb{R}_+$ of product durability. This function is supposed to be increasing and convex, i.e. $\kappa'(\phi), \kappa''(\phi) > 0$ for all $\phi \in \mathbb{R}_+$. In the second period, unit costs are $\bar{\kappa} \in \mathbb{R}_+$ which is exogenously given (which one might interpret as implying constant durability in period 2).

The decay process of firm i's first-period output is described by the two-period decay function (2.10) introduced on p. 22. Hence, ϕ_i is the share of firm i's first-period output which is still available for use in the second period and $\phi_i y_{1i}$ represents those units of firm i's first-period output which carry over to period 2. Firm i's stock of the durable good is given by (4.42). For simplicity, the initial stock is assumed to be zero. Firm i's first-period stock c_{1i} is therefore identical to first-period output y_{1i} whereas firm i's second-period stock c_{2i} consists of second period output y_{2i} and the remaining first-period units $\phi_i y_{1i}$. If firms sell their output, then the products are the property of consumers and c_{ti} should rather be designated as that part of the consumers' stock of the durable good in period t which stems from production of firm i. To avoid clumsy phrases, however, we will continue to refer to c_{ti} as firm i's stock of the durable good in case of both sales and renting.

---

[8] The analysis in this section is based on Runkel (1999, 2003a).

p_t in (4.43) is the rental price of the durable good or, equivalently, the price of the durable's services in period t as a function of the total industry stock C_t. The demand function P is supposed to satisfy assumption A4 on p. 25. In addition, we assume

$$P'(Y_2 + \bar{\phi}Y_1) + y_{2i}P''(Y_2 + \bar{\phi}Y_1) < 0 \quad \forall\, Y_2 + \bar{\phi}Y_1 \in \mathbb{R}_+, \quad i \in N , \quad (4.46)$$

where $Y_2 = \sum_{j \in N} y_{2j}$ is the industry output in period 2 and $\bar{\phi}Y_1 = \sum_{j \in N} \phi_j y_{1j}$ denotes the industry amount of first-period units still in use in the second period. (4.46) is the usual stability condition in oligopoly models applied to second-period sales revenues. It states that firm i's marginal sales revenues in period 2 are decreasing in the total industry stock $Y_2 + \bar{\phi}Y_1 = C_2$ in period 2.

It remains to specify how consumption waste is modeled. This is done in (4.44) and (4.45). Analogous to Sect. 4.1, we assume all units of the durable good to be of constant weight regardless of their product durability. As a consequence, the amount of waste is equal to the amount of scrapped units. In period 1, $(1 - \phi_i)y_{1i}$ units of firm i's first-period output wear out such that firm i's waste in period 1 is $w_{1i} = (1 - \phi_i)c_{1i}$. In period 2, all units are assumed to be scrapped with the consequence that firm i's waste in period 2 reads $w_{2i} = c_{2i}$. The industry amount of waste in period t is W_t defined in (4.45). It gives rise to external disposal and environmental costs resulting from waste transport, incineration, recycling and landfilling. The external costs at time t are expressed by the damage function $E(W_t)$ with $E' > 0$ and $E'' \geq 0$, i.e. waste causes damage at non-decreasing rates.

Like in the vintage model of Sect. 4.1, product durability has a waste-delaying and a waste-reducing effect in the two-period model (4.41)–(4.45). To identify the waste-delaying effect, rearrange (4.42) and (4.44) to write $w_{1i} = (1 - \phi_i)y_{1i}$ and $w_{2i} = y_{2i} + \phi_i y_{1i}$, totally differentiate these terms and set $dy_{1i} = dy_{2i} = 0$. The result is

$$\left. \frac{dw_{1i}}{d\phi_i} \right|_{\substack{dy_{1i}=0 \\ dy_{2i}=0}} = -y_{1i} < 0 \quad \text{and} \quad \left. \frac{dw_{2i}}{d\phi_i} \right|_{\substack{dy_{1i}=0 \\ dy_{2i}=0}} = y_{1i} > 0 .$$

For constant output, an increase in product durability reduces first-period waste, but raises second-period waste to the same extent. Hence, keeping output of the durable good constant and making the first period units more durable merely postpones waste generation from the first to the second period, but leaves unaltered the total waste generated by first-period output. The waste-reducing effect of durability is revealed by totally differentiating (4.42) and (4.44) and setting $dc_{1i} = dc_{2i} = 0$:

$$\left. \frac{dw_{1i}}{d\phi_i} \right|_{\substack{dc_{1i}=0 \\ dc_{2i}=0}} = -c_{1i} < 0 \quad \text{and} \quad \left. \frac{dw_{2i}}{d\phi_i} \right|_{\substack{dc_{1i}=0 \\ dc_{2i}=0}} = 0 .$$

If the stock of the durable good is held constant and if product durability is increased, then first-period waste declines while second-period waste remains unchanged. The reason for the waste reduction in period 1 is that output and product durability are substitutes with respect to the durable stock and that increasing durability allows reducing output and hence waste. To avoid a clumsy presentation, the subsequent analysis focuses on the waste-reducing effect of product durability only.

To characterize the social welfare optimum, we assume the social planner seeks a symmetric solution $x_{ti} = x_t$ and $\phi_i = \phi$ for all $x \in \{y, c, w\}$, $i \in N$ and $t \in \{1, 2\}$. The welfare maximization problem, expressed in terms of the durable stocks c_1, c_2 and durability ϕ, reads

$$\max_{c_1, c_2, \phi} V = S(nc_1) - n\kappa(\phi)c_1 - E[n(1 - \phi)c_1]$$

$$+ \rho\Big[S(nc_2) - n\bar{\kappa}(c_2 - \phi c_1) - E(nc_2)\Big], \quad (4.47)$$

where $S(C) = \int_0^C P(\xi)d\xi$ represents the consumers' benefits from consuming the services of the durable good and where $\rho = 1/(1 + \delta) \in]0, 1[$ is the social discount factor. In both periods, social welfare equals consumers' benefits less production and disposal costs. (4.47) states that the social planner maximizes the present value of social welfare. The FOCs for solving (4.47) are

$$P(nc_1^o) + \rho\bar{\kappa}\phi^o = \kappa(\phi^o) + (1 - \phi^o)E'[n(1 - \phi^o)c_1^o], \quad (4.48)$$

$$P(nc_2^o) = \bar{\kappa} + E'(nc_2^o), \quad (4.49)$$

$$\kappa'(\phi^o) = \rho\bar{\kappa} + E'[n(1 - \phi^o)c_1^o], \quad (4.50)$$

where c_1^o, c_2^o and ϕ^o denote the efficient values of the durable stock in period 1 and 2 and of product durability, respectively. The SOCs of welfare maximization are satisfied since $P' < 0$, $E'' \geq 0$ and $\kappa'' > 0$. The FOCs (4.48)–(4.50) are again of the type 'marginal social costs equal to marginal social benefits'. For a marginal increase in the first-period stock of the durable good, the LHS of (4.48) gives marginal social benefits which equal marginal consumer benefits from both periods, and the RHS of (4.48) represents marginal social costs which consist of marginal production costs from period 1 and marginal disposal costs from both periods.[9] Similarly, equation (4.49) equates marginal social benefits and marginal social costs of the second-period stock of the durable good. For a marginal increase in product durability, the LHS of (4.50) equals marginal social costs consisting of marginal production costs whereas the RHS of (4.50) represents marginal social benefits. These benefits comprise the production costs saved in period 2 and the disposal costs

---

[9] For this interpretation, solve (4.49) with respect to $\bar{\kappa}$ and insert the resulting expression into (4.48).

saved in period 1 due to the waste-reducing effect of product durability. In sum, the efficiency conditions (4.48)–(4.50) are interpreted in the same way as the efficiency conditions (4.18) and (4.21) in the vintage model of Sect. 4.1. Consequently, by carrying out a comparative static analysis of the efficiency conditions (4.48)–(4.50) it is straightforward to show that marginal disposal costs and efficient product durability are positively correlated. For the corresponding result in the vintage model, see Proposition 4.1. We refrain from repeating a formal proof of this assertion here.

4.2.2 Market Equilibrium and Laissez Faire

Now suppose the durable good is supplied by profit-maximizing firms. As mentioned already in Sect. 2.2.1 on p. 35 and as formally shown by Bulow (1986) in a durable good model without environmental pollution, the lack of precommitment ability of producers renders rental and sales equilibria different. We therefore analyze both cases separately and begin with the rental case.

Rental Equilibrium

In the rental case, the firm remains the owner of the units it produces. Rather than selling the good, the producer sells the services of the good. The present value of firm i's rental profits may be written as

$$\Pi^{ri} = \Pi_1^{ri} + \rho\Pi_2^{ri} \, , \tag{4.51}$$

where

$$\Pi_1^{ri} = c_{1i}P(C_1) - c_{1i}\Big[\kappa(\phi_i) + (1 - \phi_i)\tau_{w_1} + \tau_{c_1} + \tau_{y_1}\Big] - \phi_i\tau_\phi \, , \tag{4.52}$$

$$\Pi_2^{ri} = c_{2i}P(C_2) - (c_{2i} - \phi_i c_{1i})(\bar\kappa + \tau_{y_2}) - c_{2i}(\tau_{w_2} + \tau_{c_2}) \tag{4.53}$$

are rental profits in period 1 and 2, respectively. In both periods, rental profits equal rental revenues less production costs and tax payments. As in Sect. 4.1, the regulator introduces four tax rates: τ_{w_t} on solid waste in period t, τ_{y_t} on output in period t, τ_{c_t} on the stock of the durable good in period t and τ_ϕ on product durability in period 1. Renting producers own the stock of the durable good and dispose of worn-out products so that all tax rates directly enter the profit functions.

Firm i's decision variables are c_{1i}, c_{2i} and ϕ_i. It sets these variables as to maximize profits. Due to the lack of commitment ability, the firm does not precommit at the beginning of the period 1, but recalculates its initial decision at the beginning of period 2. Hence, the durable good firm without commitment ability employs closed-loop rather than open-loop strategies. Under a closed-loop information structure, firm i takes into account that its current actions influence its own and the rivals' future actions. In order to properly calculate this influence, firm i solves the profit maximization problem recursively. It begins in period 2 and then turns to period 1 where it accounts for the future effects of its decision. The definition of a closed-loop equilibrium of the durable good industry is almost the same as definition D2 of an open-loop equilibrium presented in Sect. 2.2.1. The only difference is that firms now use closed-loop rather than open-loop strategies. To derive the closed-loop equilibrium, we therefore first derive the Nash-equilibrium in the second-period subgame where all first-period variables are taken as given and then turn to the equilibrium in period 1 where the influence of first-period variables on the second-period equilibrium is taken into account.

In the second period, firm i maximizes (4.53) with respect to c_{2i} taking as given c_{1i}, ϕ_i and c_{2j} for all $j \neq i$. The FOC reads

$$P(C_2^r) + c_{2i}^r P'(C_2^r) = \bar{\kappa} + \tau_{w_2} + \tau_{y_2} + \tau_{c_2} \qquad \forall i \in N , \qquad (4.54)$$

where the superscript r indicates the rental case. The SOC of the firm's second-period profit maximum is satisfied by assumption A4. To see that (4.54) determines an unique and stable equilibrium of the rental subgame in period 2, observe that, for given $\phi_i c_{1i}$, maximization of (4.53) is a special case of profit maximization in the nondurable good oligopoly considered in Dixit (1986) since we may rewrite (4.53) as $\Pi_2^{ri} = c_{2i} P(C_2) - \widetilde{K}(c_{2i})$ with the linear cost function $\widetilde{K}(c_{2i}) = (\bar{\kappa} + \tau_{w_2} + \tau_{y_2} + \tau_{c_2})c_{2i} - (\bar{\kappa} + \tau_{y_2})\phi_i c_{1i}$. Dixit (1986) shows that condition (2.19) in assumption A4 implies uniqueness and stability of the market equilibrium. Note that the equilibrium second-period stocks do not depend on the first-period units still available in period 2 since the equilibrium conditions (4.54) obviously imply $\partial c_{2j}^r / \partial \phi_i c_{1i} = 0$ for all $j, i \in N$. Thus, first-period decisions in rental markets do not influence the second-period equilibrium. This *result* of our analysis of the closed-loop rental equilibrium coincides with the *assumption* imposed in case of an open-loop information structure. Therefore, it is already clear that open-loop and closed-loop rental equilibria are the same.

In period 1, firm i maximizes the present value of rental profits in equation (4.51) with respect to c_{1i} and ϕ_i. It takes as given ϕ_j and c_{1j} of all competitors $j \neq i$ and, in general, accounts for the influence of c_{1i} and ϕ_i on c_{2j}^r for all $j \in N$ since we suppose closed-loop strategies. But from the analysis of second-period profit maximization we know $\partial c_{2j}^r / \partial \phi_i c_{1i} = 0$ for all $j, i \in N$ which amounts to firm i taking as given c_{2j}^r for all $j \in N$ in period 1. The FOCs of firm i's profit maximum are then obtained by setting equal to zero the partial derivatives of (4.51) with respect to c_{1i} and ϕ_i. The resulting expressions are

$$P(C_1^r) + c_{1i}^r P'(C_1^r) + \rho\bar{\kappa}\phi_i^r = \kappa(\phi_i^r) + (1 - \phi_i^r)\tau_{w_1}$$

$$+ \tau_{y_1} - \rho\phi_i^r \tau_{y_2} + \tau_{c_1} \qquad \forall i \in N , \qquad (4.55)$$

$$\kappa'(\phi_i^r) + \tau_\phi / c_{1i}^r = \rho\bar{\kappa} + \rho\tau_{y_2} + \tau_{w_1} \qquad \forall i \in N . \qquad (4.56)$$

It can be shown that the SOCs of first-period maximization are fulfilled if and only if the absolute value of the durability tax rate satisfies

$$|\tau_\phi| \leq \sqrt{-c_{1i}^{r3}\kappa''(\phi_i^r)[P(C_1^r) + c_{1i}^r P''(C_1^r)]} \,. \tag{4.57}$$

The subsequent analysis proceeds on the assumption that τ_ϕ satisfies (4.57). An example is the case where durability taxation is absent $(\tau_\phi = 0)$.

Summing up, conditions (4.54)–(4.56) characterize the market equilibrium in rental markets without commitment ability of producers. (4.54) and (4.55) say that marginal revenues of the durable stock in both periods equal the respective marginal production costs plus marginal tax payments. According to (4.56), marginal production costs and marginal tax payments of increasing durability are offset by the production costs saved in period 2, the output tax payments saved in period 2 and the waste tax payments saved in period 1. As already mentioned above, these conditions for a closed-loop rental equilibrium are the same as for an open-loop rental equilibrium where the open-loop conditions are obtained by directly differentiating (4.51) with respect to c_{1i}, c_{2i} and ϕ_i. In the subsequent analysis, we therefore can use (4.54)–(4.56) not only as the conditions for a closed-loop rental equilibrium, but also for an open-loop rental equilibrium.

Before turning to the sales case, let us check whether Swan's independence result holds in the rental equilibrium determined by (4.54)–(4.56). We focus on a symmetric equilibrium $\phi_i^r = \phi^r$ and $x_{ti}^r = x_t^r$ for all $x \in \{y, c\}$ and $t \in \{1, 2\}$, totally differentiate the equilibrium conditions (4.54)–(4.56) and apply Cramer's rule to obtain

$$\frac{\partial \phi^r}{\partial n} = -\frac{1}{|J^r|}(P_1' + c_1^r P_1'')[(n+1)P_2' + nc_2^r P_2'']\frac{\tau_\phi}{c_1^r} \,, \tag{4.58}$$

where $P_t' := P'(nc_t^r)$ and $P_t'' := P''(nc_t^r)$ for $t \in \{1, 2\}$. The Jacobian determinant $|J^r| = [(n+1)P_2' + nc_2^r P_2'']\{\kappa''[(n+1)P_1' + nc_1^r P_1''] + \tau_\phi^2/c_1^{r3}\}$ is positive due to assumption A4 and (4.57). $|J^r| > 0$ together with assumption A4 ensures the three first terms in (4.58) are positive. Hence, the sign of $\partial \phi^r/\partial n$ equals the sign of $-\tau_\phi$. We have proved

Proposition 4.7. *In the rental equilibrium*

$$\frac{\partial \phi^r}{\partial n} \left\{\begin{array}{c} > \\ = \\ < \end{array}\right\} 0 \qquad \Leftrightarrow \qquad \tau_\phi \left\{\begin{array}{c} < \\ = \\ > \end{array}\right\} 0 \,.$$

The intuition behind this result becomes clear by considering the equilibrium condition (4.56). For the time being, set all tax rates equal to zero except for the durability tax. If τ_ϕ is negative, then marginal costs of increasing durability consist of marginal production costs κ', and marginal benefits comprise production costs $\rho\bar{\kappa}$ saved in period 2 as well as the durability subsidy *per unit* of the first-period stock, $-\tau_\phi/c_1^r$. An increase in the number of firms reduces the individual stock c_1^r and, thus, raises marginal benefits of durability by increasing the subsidy per unit. This effect provides the renting firms with the incentive to raise their product durability upon an increase in the number of competitors ($\partial\phi^r/\partial n > 0$ if $\tau_\phi < 0$). Hence, if durability is subsidized, then the monopolist chooses the smallest durability. By the same argument, the monopolist provides the greatest durability if durability is taxed ($\partial\phi^r/\partial n < 0$ if $\tau_\phi > 0$). Swan's independence result remains valid only for the special case where the durability tax rate is equal to zero.

One may wonder why the validity of Swan's result is influenced by durability taxation only. In the vintage model of Sect. 4.1, output and waste taxation also upset the independence result as shown by equation (4.33). The reason for this difference is that in the vintage model, we assumed a cost function satisfying the Abel condition while in the two-period model, we focus on a linear cost function. The Abel condition is a rather general condition which implies that the cost-minimizing product durability is independent of the firms' market share only when the tax rates on output, durability and waste are zero; if one of these tax rates is non-zero, then the independence result no longer holds. In contrast, the assumption of linear costs is more restrictive than the Abel condition and therefore leaves more leeway for the validity of the independence result, i.e. introducing waste and/or output taxation under a linear cost function does not upset the independence result. This observation reveals that it is always the *combination* of costs conditions and tax payments which determine whether Swan's result is valid or not.

Sales Equilibrium

Now turn to the case in which producers sell the durable good instead of selling the durable's services. In the absence of all taxes, the sales price of the durable good equals the present value of service prices (implicit rentals) which the durable generates over its whole lifetime. Firm i's sales revenues from first-period and second-period output then read $y_{1i}[P(Y_1) + \rho\phi_i P(Y_2 + \bar{\phi}Y_1)]$ and $y_{2i}P(Y_2 + \bar{\phi}Y_1)$, respectively. Now we introduce the same taxes as in the analysis of rental markets in the previous subsection. Waste and stock taxes are paid by consumers since in sales markets the products are in the consumers' property. Their stock tax payments for units coming from firm i's first-period output are $y_{1i}\tau_{c_1}$ in period 1 and $\rho\phi_i y_{1i}\tau_{c_2}$ in period 2. Consumers' waste tax payments for units coming from firm i's first-period output are $(1-\phi_i)y_{1i}\tau_{w_1}$ in period 1 and $\rho\phi_i y_{1i}\tau_{w_2}$ in period 2. When purchasing the durable good, rational utility-maximizing consumers anticipate these tax payments and adjust their willingness-to-pay for the products appropriately. Consequently, the introduction of waste and stock taxes reduces firm i's sales revenues from first-period output to $y_{1i}[P(Y_1)+\rho\phi_i P(Y_2+\bar{\phi}Y_1)-(1-\phi_i)\tau_{w_1}-\tau_{c_1}-\rho\phi_i(\tau_{w_2}+\tau_{c_2})]$. By the same arguments, firm i's revenues from second-period output become $y_{2i}[P(Y_2+\bar{\phi}Y_1)-\tau_{w_2}-\tau_{c_2}]$ since consumers spend $y_{2i}\tau_{w_2}$ and $y_{2i}\tau_{c_2}$ on waste and stock tax payments for units stemming from firm i's second-period output.[10] Moreover, if we also consider output and durability taxes (which are usually paid by producers) and production costs of the durable good, then firm i's present value of sales profits may be written as

$$\Pi^{si} = \tilde{\Pi}_1^{si} + \rho\tilde{\Pi}_2^{si} ,$$

where

$$\tilde{\Pi}_1^{si} = y_{1i}\Big[P(Y_1) + \rho\phi_i P(Y_2 + \bar{\phi}Y_1) - (1 - \phi_i)\tau_{w_1} - \tau_{c_1} - \tau_{y_1}$$

$$-\rho\phi_i(\tau_{w_2} + \tau_{c_2}) - \kappa(\phi_i)\Big] - \tau_\phi\phi_i ,$$

$$\tilde{\Pi}_2^{si} = y_{2i}\Big[P(Y_2 + \bar{\phi}Y_1) - \tau_{w_2} - \tau_{c_2} - \tau_{y_2} - \bar{\kappa}\Big]$$

are firm i's sales profits in period 1 and 2, respectively.

Firm i's decision variables are first-period and second-period output, y_{1i} and y_{2i}, and product durability, ϕ_i. Once y_{1i}, y_{2i} and ϕ_i are determined, so are firm i's stocks of the durable good $c_{1i} = y_{1i}$ and $c_{2i} = y_{2i} + \phi_i y_{1i}$. It therefore does not matter whether we regard $\{y_{1i}, y_{2i}, \phi_i\}$ or $\{c_{1i}, c_{2i}, \phi_i\}$ as firm i's pertinent decision variables. To be consistent with the analysis of rental markets, however, it proves useful to regard firm i as choosing c_{1i}, c_{2i} and ϕ_i. We therefore use the equations $c_{1i} = y_{1i}$ and $c_{2i} = y_{2i} + \phi_i y_{1i}$ in order to rewrite Π^{si} as

$$\Pi^{si} = \Pi_1^{si} + \rho\Pi_2^{si} , \tag{4.59}$$

where

$$\Pi_1^{si} = c_{1i}\Big[P(C_1) + \rho\phi_i P(C_2) - (1 - \phi_i)\tau_{w_1} - \tau_{c_1} - \tau_{y_1}$$

$$-\rho\phi_i(\tau_{w_2} + \tau_{c_2}) - \kappa(\phi_i)\Big] - \tau_\phi\phi_i , \tag{4.60}$$

$$\Pi_2^{si} = (c_{2i} - \phi_i c_{1i})\Big[P(C_2) - \tau_{w_2} - \tau_{c_2} - \tau_{y_2} - \bar{\kappa}\Big] . \tag{4.61}$$

If firm i possesses commitment ability, then it uses open-loop strategies in maximizing profits. The solution to this profit maximization problem is obtained by directly differentiating (4.59) with respect to firm i's decision variables. It is straightforward to show that the resulting equilibrium conditions are equal to the equilibrium conditions (4.54)–(4.56) in rental markets. Hence,

---

[10] As in the vintage model considered in Sect. 4.1, the consumer demand referred to in this argumentation can be shown to result from the (two-period) utility maximization of a representative consumer who is charged with waste and stock taxes.

(4.54)–(4.56) determine not only the the open-loop and closed-loop rental equilibria, but also the open-loop sales equilibrium. This observation shows that in the presence of commitment ability, the equivalence of sales and rental equilibria derived in Proposition 2.2 is also valid in the two-period model considered in the present section.

Let us now consider the case where durable good producers are not able to commit, which is our principle focus in this section. Assuming firms to use closed-loop strategies means that they take into account the influence of their current decisions on its own and the rivals' future actions. Each firm determines this influence by solving the profit maximization problem recursively. In period 2, the selling firm i regards the first-period as water under the bridge and maximizes (4.61) with respect to c_{2i} taking as given not only the second-period stock c_{2j} of all other firms $j \neq i$, but also its own first-period units $\phi_i c_{1i}$ still in use in period 2. The pertinent FOC reads

$$P(C_2^s) + (c_{2i}^s - \phi_i c_{1i})P'(C_2^s) = \bar{\kappa} + \tau_{w_2} + \tau_{y_2} + \tau_{c_2} \qquad \forall i \in N , \quad (4.62)$$

where the superscript s indicates the sales case. The SOC is satisfied due to (4.46). (4.62) implicitly determines an unique and stable equilibrium of the second-period sales subgame when combined with condition (4.46).[11] The equilibrium stocks in period 2 are functions of the remaining first-period units, i.e. $c_{2j}^s = c_{2j}^s(\phi \mathbf{c_1})$ for all $j \in N$. Applying the comparative static methods of Dixit (1986) to (4.62) yields

$$\frac{\partial c_{2i}^s}{\partial \phi_i c_{1i}} = \frac{nP_2' + (Y_2^s - y_{2i}^s)P_2''}{(n+1)P_2' + Y_2^s P_2''} \in]0,1[\qquad \forall i \in N , \quad (4.63)$$

$$\frac{\partial c_{2j}^s}{\partial \phi_i c_{1i}} = -\frac{P_2' + y_{2j}^s P_2''}{(n+1)P_2' + Y_2^s P_2''} \in]-1,0[\qquad \forall j,i \in N, j \neq i . \quad (4.64)$$

For $C_2^i := \sum_{j \neq i} c_{2j}$, equation (4.64) implies

$$\frac{\partial C_2^{i,s}}{\partial \phi_i c_{1i}} = -\frac{(n-1)P_2' + (Y_2^s - y_{2i}^s)P_2''}{(n+1)P_2' + Y_2^s P_2''} \in]-1,0] \qquad \forall i \in N . \quad (4.65)$$

The signs of (4.63)–(4.65) follow from (4.46) and $P_2' < 0$. They have a straightforward interpretation: The greater firm i's first-period output $y_{1i} = c_{1i}$ and/or the greater firm i's product durability ϕ_i, the greater is ceteris paribus firm i's own equilibrium second-period stock of the durable good and the lower

---

[11] This assertion is most easily verified by considering the 'untransformed' problem of maximizing $\tilde{\Pi}_2^{si}$ with respect to y_{2i}. The FOCs $P(Y_2^s + \bar{\phi}Y_1) + y_{2i}^s P'(Y_2^s + \bar{\phi}Y_1) = \bar{\kappa} + \tau_{w_2} + \tau_{y_2} + \tau_{c_2}$ for all $i \in N$ are equivalent to (4.62). Furthermore, for given $\bar{\phi}Y_1$, the untransformed problem is again a special case of profit-maximization in the nondurable good oligopoly considered by Dixit (1986). Consequently, condition (4.46) implies uniqueness and stability in the second-period sales subgame under consideration.

are the equilibrium second-period stocks of firm i's competitors. Hence, in contrast the rental markets, in sales markets, firm i is able to employ first-period output and durability as strategic variables to influence the second-period equilibrium in its own favor.

This influence is taken into account by firm i when making its decision in the first period. In period 1, firm i chooses c_{1i} and ϕ_i such as to maximize the present value of its profits subject to $c_{2j}^s = c_{2j}^s(\phi c_1)$. Rearranging (4.59)–(4.61), firm i's first-period problem may be written as

$$
\max_{c_{1i},\phi_i} \quad \Pi^{si} = c_{1i}\left[P\left(\sum_{j\in N} c_{1j}\right) - (1-\phi_i)\tau_{w_1} - \tau_{y_1} - \tau_{c_1} - \kappa(\phi_i)\right] - \tau_\phi \phi_i
$$

$$
+ \rho\left[c_{2i}^s(\phi c_1)\left[P\left(\sum_{j\in N} c_{2j}^s(\phi c_1)\right) - \tau_{w_2} - \tau_{y_2} - \tau_{c_2} - \bar\kappa\right]\right.
$$

$$
\left. + \phi_i c_{1i}(\tau_{y_2} + \bar\kappa)\right] .
$$

Using (4.62) for simplifying terms, the pertinent FOCs turn out to be[12]

$$
P(C_1^s) + c_{1i}^s P'(C_1^s) + \rho\bar\kappa\phi_i^s = \kappa(\phi_i^s) + (1 - \phi_i^s)\tau_{w_1}
$$

$$
+ \tau_{y_1} + \tau_{c_1} - \rho\phi_i^s\tau_{y_2} - \rho\phi_i^s z_i \qquad \forall\, i \in N \ , \quad (4.66)
$$

$$
\kappa'(\phi_i^s) + \tau_\phi/c_{1i}^s = \rho\bar\kappa + \tau_{w_1} + \rho\tau_{y_2} + \rho z_i \qquad \forall\, i \in N \ , \quad (4.67)
$$

with

$$
z_i := \underbrace{\phi_i^s c_{1i}^s P'(C_2^s)\frac{\partial c_{2i}^s}{\partial \phi_i^s c_{1i}^s}}_{=:z_i^- < 0} + \underbrace{c_{2i}^s P'(C_2^s)\frac{\partial C_2^{i,s}}{\partial \phi_i^s c_{1i}^s}}_{=:z_i^+ > 0} \qquad \forall\, i \in N \ . \quad (4.68)
$$

The signs of z_i^- and z_+^i in (4.68) follow from (4.63) and (4.65). Equations (4.62), (4.66) and (4.67) combined with (4.63), (4.65) and (4.68) represent the conditions for a closed-loop sales equilibrium.

These equilibrium conditions reveal several reasons why the laissez faire equilibrium allocation of a selling durable good oligopoly with endogenous product durability and without commitment ability is inefficient even in the absence of external disposal costs. For convenience of later reference, we designate the distortions as D1 to D5:

---

[12] Due to the complexity of the first-period profit maximization problem, I did not succeed in finding general conditions under which the SOCs are satisfied. Nevertheless, in the proofs of Propositions 4.15 and 4.16 in Sect. 4.2.4, two numerical examples are presented where the SOCs are proved to hold. This ensures that our analysis of the sales equilibrium without precommitment is not empty.

- Distortion D1: The terms $c_{1i}^s P'(C_1^s)$ in (4.66) and $(c_{2i}^s - \phi_i^s c_{1i}^s) P'(C_2^s)$ in (4.62) represent the conventional distortion due to the firms' market power. As in the textbook model of imperfect competition, these terms tend to cause an underprovision of output in both periods, ceteris paribus.

- Distortion D2: Through its negative component z_i^-, the term z_i in (4.66) diverts first-period output in the same direction as D1: All other things being equal, firm i benefits from reducing $y_{1i} = c_{1i}$ since this stimulates second-period sales $y_{2i} = c_{2i} - \phi_i c_{1i}$. To see this, differentiate y_{2i} with respect to c_{1i} and use (4.63). Hence, D2 induces firm i to reduce first-period output below the efficient level, ceteris paribus.

- Distortion D3: The positive component z_i^+ of z_i in (4.66) works in the opposite direction as D1 and D2. The intuition is that firm i, ceteris paribus, profits from raising first-period output since this absorbs the consumers' future willingness-to-pay already in period 1 with the consequence that i's own revenues increase while its rivals' revenues shrink due to (4.64) and (4.65). In the subsequent analysis, we refer to this distortion as firm i's *revenue raising* incentive.

- Distortion D4: The negative component z_i^- of z_i in (4.67) has similar implications as in (4.66). Firm i is prone to reduce durability below the efficient level since second-period sales $y_{2i} = c_{2i} - \phi_i c_{1i}$ are enhanced not only through a lower first-period output $y_{1i} = c_{1i}$, but also through a smaller durability ϕ_i. This can be seen by differentiating y_{2i} with respect to ϕ_i and using (4.63). This effect represents the firm's incentive for *planned obsolescence* which we already referred to in the introduction on p. 5.

- Distortion D5: The positive part z_i^+ of z_i in (4.67) has an analogous interpretation as in (4.66). Firm i increases product durability in order to snatch away consumers' future willingness-to-pay from competitors. This behavior improves i's own revenues while it reduces the rivals' revenues due to (4.64) and (4.65).[13]

In the general case of durable good oligopoly with endogenous product durability, distortions D1 to D5 are all present with the consequence that z_i may take any sign in the market equilibrium conditions. The subsequent analysis will not only consider this general case, but also the special cases 2–4 listed in Table 4.2. In case 2, durability is not a choice variable of durable good sellers with the consequence that producers do not have an incentive for planned obsolescence (D4 absent) and cannot use durability as a measure to absorb future willingness-to-pay of consumers (D5 absent). Formally, this follows from the fact that in case of exogenous product durability, the equilibrium condition (4.67) does not apply. z_i then still shows up in the equilibrium

---

[13] Distortion D5 represents the revenue raising effect of durability which we already referred to in the introduction on p. 5. In order to avoid confusion with distortion D3, we do not designate D5 as revenues raising effect, however.

Table 4.2. Distortions of the Market Equilibrium

case		D1	D2	D3	D4	D5	z_i
1	selling oligopoly with endogenous durability	+	+	+	+	+	$\gtreqless 0$
2	selling oligopoly with exogenous durability	+	+	+	–	–	$\gtreqless 0$
3	selling monopoly with endogenous durability	+	+	–	+	–	< 0
4	selling monopoly with exogenous durability	+	+	–	–	–	< 0
5	rental markets under imperfect competition	+	–	–	–	–	$= 0$
6	perfect competition (sales and rental case)	–	–	–	–	–	$= 0$

(+ = distortion is present; – = distortion is not present)

condition (4.66) and is still indeterminate in sign. Case 3 in Table 4.2 is a selling durable good monopolist with endogenous durability. This monopolist faces no competition and, thus, has no incentive to snatch away from competitors the future willingness-to-pay of consumers. Consequently, distortions D3 and D5 are absent. Formally, this follows from equation (4.65) that implies $\partial C_2^{i,s}/\partial \phi_i c_{1i} = 0$ for $n = 1$ and hence $z_i = z_i^- < 0$ via equation (4.68). Case 4 of Table 4.2 combines the conditions of the cases 2 and 3. It represents a durable goods monopoly with given product durability so that distortions D3, D4 and D5 are not present and z_i becomes negative.

The rental equilibrium conditions (4.54)–(4.56) differ from the sales equilibrium conditions (4.62), (4.66) and (4.67) only with respect to the terms containing z_i. We can therefore view the rental equilibrium also as 'special case' of the sales equilibrium. In this special case, z_i is zero and the distortions D2–D5 are absent (case 5 in Table 4.2). Furthermore, also in this section we will consider the case of perfect competition which is identical to case 6 in Table 4.2. Under perfect competition, P' is equal to zero with the consequence that all distortions D1–D5 vanish in both rental and sales markets. In the subsequent analysis of environmental policy, it will be of great importance to know which distortions are at work in the durable good industry.

Laissez Faire Equilibrium

Before turning to this policy analysis, we make the case for governmental regulation by identifying market failure under laissez faire. For simplicity, from now on we focus on symmetric equilibria. In the rental case, the equilibrium values are $\phi_i^r = \phi^r$ and $x_{ti}^r = x_t^r$ for all $i \in N$, $x \in \{y, c, w\}$ and $t \in \{1, 2\}$. The equilibrium values in the sales case are denoted by $\phi_i^s = \phi^s$ and $x_{ti}^s = x_t^s$ for all $i \in N$, $x \in \{y, c, w\}$ and $t \in \{1, 2\}$. It then follows

Proposition 4.8. *Suppose* $\tau_{w_t} = \tau_{y_t} = \tau_{c_t} = \tau_\phi = 0$ *for* $t \in \{1, 2\}$. *Then,*
(i) the market equilibrium under perfect competition satisfies

$$c_t^s = c_t^r > c_t^o , \qquad w_t^s = w_t^r > w_t^o \qquad \forall t \in \{1, 2\} ,$$

$$\phi^s = \phi^r < \phi^o , \qquad y_1^s = y_1^r > y_1^o , \qquad y_2^s = y_2^r \gtreqqless y_2^o ,$$

(ii) in the market equilibrium under imperfect competition

$$n = 1 \quad implies \quad \phi^s < \phi^r < \phi^o$$

whereas

$$n > 1 \quad implies \quad \phi^r < \phi^o$$

$$and \qquad \phi^s \gtreqqless \phi^r \quad \Leftrightarrow \quad z \gtreqqless 0$$

$$and \qquad \phi^s \gtreqqless \phi^o \quad \Leftrightarrow \quad z \gtreqqless \frac{1}{\rho} E'[n(1 - \phi^o)c_1^o] .$$

For all other variables, the comparison of efficient and equilibrium values under imperfect competition is ambiguous, in general.

Proof. For notational convenience, define

$$F(\phi) := \kappa(\phi) - \rho\bar{\kappa}\phi \quad with \quad F'(\phi) := \kappa'(\phi) - \rho\bar{\kappa} , \quad F''(\phi) := \kappa''(\phi) > 0 ,$$

and

$$G(nc_2) := P(nc_2) - \bar{\kappa} \quad with \quad G'(nc_2) = P'(nc_2) < 0 .$$

Under perfect competition, we have $P' \equiv 0$. The conditions (4.54)–(4.56) for the rental equilibrium then coincide with the conditions (4.62), (4.66) and (4.67) for the sales equilibrium. Hence, equilibrium variables in the rental equilibrium are the same as in the sales equilibrium and we can restrict our attention on the rental equilibrium determined by (4.54)–(4.56). According to (4.50) and (4.56), efficient durability is determined by $F'(\phi^o) = E'(\cdot) > 0$ whereas equilibrium durability satisfies $F'(\phi^r) = 0$. $F'' > 0$ then implies $\phi^s = \phi^r < \phi^o$. c_2^o and $c_2^r = c_2^s$ follow from (4.49) and (4.54), respectively. So, we obtain $G(nc_2^o) = E'(\cdot) > 0$ and $G(nc_2^r) = 0$. It follows $c_2^s = c_2^r > c_2^o$ owing to $G' < 0$. To prove $c_1^s = c_1^r > c_1^o$, note that $\phi^r < \phi^o$, $F'(\phi^o) = E'(\cdot) > 0 = F'(\phi^r)$ and $F'' > 0$ imply $F(\phi^o) > F(\phi^r)$. According to (4.48) and (4.55), $c_1^r = c_1^s$ and c_1^o are determined by $P(nc_1^o) = F(\phi^o) + (1 - \phi^o)E'(\cdot) > F(\phi^r) = P(nc_1^r)$ so that we conclude $c_1^s = c_1^r > c_1^o$ from $P' < 0$. $y_1^s = y_1^r > y_1^o$ and

$w_t^s = w_t^r > w_t^o$ follow from the equations $y_1 = c_1$, $w_1 = (1-\phi)c_1$ and $w_2 = c_2$. The proof of Proposition 4.8i is completed by accounting for $y_2^s = c_2^s - \phi^s c_1^s$ which implies $y_2^s = y_2^r \gtreqless y_2^o$.

In order to prove Proposition 4.8ii, note first that sales and rental equilibria diverge under imperfect competition. Equations (4.56) and (4.67) imply that durability in the rental equilibrium satisfies $F'(\phi^r) = 0$ whereas durability in the sales equilibrium is determined by $F'(\phi^s) = \rho z$. According to (4.50), efficient durability follows from $F'(\phi^o) = E'(\cdot)$. In monopoly, we obtain $\rho z < 0 < E'(\cdot)$ and thus $\phi^o > \phi^r > \phi^s$ since $F'' > 0$. In oligopoly, $\phi^r < \phi^o$ follows from Proposition 4.8i and Swan's independence result (proved in Proposition 4.7). Furthermore, in oligopoly, ρz is ambiguous in sign. $z \gtreqless 0$ is therefore equivalent to $\phi^s \gtreqless \phi^r$ whereas $z \gtreqless E'(\cdot)/\rho$ is equivalent to $\phi^s \gtreqless \phi^o$ owing to $F'' > 0$. To complete the proof of Proposition 4.8ii, note that the efficiency conditions (4.48)–(4.50) contain $E'(\cdot)$ while the equilibrium conditions (4.54)–(4.56) or (4.62), (4.66) and (4.67) comprise $c_t P'(\cdot)$ and the terms containing z. $E'(\cdot)$, $c_t P'(\cdot)$ and z exert opposite effects on output, stock and solid waste of the durable good in both periods. □

The laissez faire equilibrium under perfect competition is characterized by Proposition 4.8i. In both the rental and the sales case, it is distorted only by the disposal costs externality since all other distortions are absent according to case 6 of Table 4.2. Consequently, we obtain the same results as in the model with commitment ability in Sect. 4.1. Disposal costs are external to the optimization problems of unregulated durable good producers implying that waste is chosen inefficiently high in both periods. Equilibrium output of the durable good is inefficiently high since output and waste are positively correlated, and product durability is lower than in the welfare optimum since producers ignore the waste-reducing effect of product durability. The equilibrium stock of the durable good is inefficiently high in both periods.

In the laissez faire equilibrium under imperfect competition, the lack of commitment ability renders rental and sales equilibria different (Proposition 4.8ii). Moreover, owing to the multiple distortions under imperfect competition (disposal externality and distortions D1–D5) the comparison of the efficient values and the equilibrium values is inconclusive for most variables. An exception is product durability. In unregulated rental markets under imperfect competition, equilibrium durability ϕ^r falls short of efficient durability ϕ^o independent of the number of firms operating in the industry (i.e. for $n = 1$ and $n > 1$). The reason is Swan's independence result which states that durability of rented products under imperfect competition takes on the same inefficiently low value as under perfect competition. Things are more complicated in unregulated sales markets. First, as compared to renting monopoly, product durability ϕ^s in selling monopoly is further decreased below its efficient level ϕ^o: The selling monopolist not only ignores disposal costs of her products, but also has an incentive for planned obsolescence (distortion D4). Planned obsolescence distorts product durability in the same direction as the disposal externality. Durability in selling monopoly is therefore unambiguously smaller than in renting monopoly and in the welfare optimum. Second, durability ϕ^s in selling oligopoly is additionally influenced by the firms' incentive to absorb future willingness-to-pay of consumers (distortion D5). Consequently, product durability in selling oligopoly is greater than in renting oligopoly if the distortion D5 outweighs the obsolescence effect (distortion D4) so that z in (4.67) is positive. Moreover, if the incentive for absorbing willingness-to-pay (distortion D5) compensates the incentive for planned obsolescence (distortion D4) and the disposal costs externality, then $z > E'[n(1 - \phi^o)c_1^o]/\rho$ and durability becomes even greater than in the welfare optimum. It follows that in selling durable good oligopoly without commitment ability, an overprovision of product durability is possible in spite of the waste-reducing effect of product durability. This observation is in stark contrast to Proposition 4.3 in the model with commitment ability of producers where (long run) product durability falls short of its efficient level.

4.2.3 First-Best Regulation

If the regulator aims at fully correcting for the market failure identified in Proposition 4.8, then she has to compare the efficiency conditions (4.48)–(4.50) with the market equilibrium conditions (4.54)–(4.56) and (4.62)–(4.68), respectively, in order to determine those tax rates that render the market solution socially optimal. The market equilibrium is efficient if and only if the tax rates satisfy

$$(1 - \phi^o)\tau_{w_1} + \tau_{c_1} + \tau_{y_1} - \rho\phi^o\tau_{y_2} = (1 - \phi^o)E'[n(1 - \phi^o)c_1^o]$$

$$+ c_1^o P'(nc_1^o) + \varphi\rho\phi^o z , \qquad (4.69)$$

$$\tau_{w_2} + \tau_{c_2} + \tau_{y_2} = E'(nc_2^o) + (c_2^o - \varphi\phi^o c_1^o)P'(nc_2^o) , \quad (4.70)$$

$$\tau_{w_1} + \rho\tau_{y_2} - \tau_\phi/c_1^o = E'[n(1 - \phi^o)c_1^o] - \varphi\rho z , \qquad (4.71)$$

where φ is a dummy variable with $\varphi = 0$ representing the rental case and $\varphi = 1$ representing the sales case. Since the number of tax rates exceeds the number of equations, there is a continuum of first-best tax-subsidy scheme. Unlike in the previous chapters, it is left to the reader to investigate special cases in which the regulatory scheme (4.69)–(4.71) comprises of at most two non-zero tax rates. We proceed, instead, by focusing on the implications the lack of commitment ability has for efficient regulation. This is done by means of the subsequent Propositions 4.9–4.12.

Let us start with the case of perfect competition constituted by $P' = z = 0$. The regulatory scheme (4.69)–(4.71) then leads to

Proposition 4.9. *Under perfect competition, both the rental and the sales equilibrium are efficient if*

$$\tau_{w_1} = E'[n(1 - \phi^o)c_1^o] > 0, \qquad \tau_{w_2} = E'(nc_2^o) > 0,$$

$$\tau_{c_t} = \tau_{y_t} = \tau_\phi = 0 \qquad for \quad t = 1, 2.$$

According to this proposition, Pigouvian taxation is an appropriate policy approach under perfect competition even if producers do not possess commitment ability. The reason is that distortions D1–D5 are not present under perfect competition, and the market equilibrium (both in the rental and sales case) is distorted by the disposal externality only. This externality is removed by imposing a Pigouvian tax on solid consumption waste.

Turning to the case of imperfect competition ($P' < 0$), we need to distinguish between regulation in rental and sales markets. For the rental case ($\varphi = 0$), equations (4.69)–(4.71) yield

Proposition 4.10. *Under imperfect competition, the rental equilibrium is efficient if*

$$\tau_{w_1} = E'[n(1 - \phi^o)c_1^o] > 0, \qquad \tau_{w_2} = E'(nc_2^o) > 0,$$

$$\tau_{c_1} = c_1^o P'(nc_1^o) < 0, \qquad \tau_{c_2} = c_2^o P'(nc_2^o) < 0,$$

$$\tau_{y_t} = \tau_\phi = 0 \qquad for \quad t \in \{1, 2\}.$$

The equilibrium in rental markets without commitment ability of producers is distorted only by the disposal externality and the usual market imperfection due to market power of firms (distortion D1). As shown in case 5 of Table 4.2, the other distortions D2–D5 are absent. Consequently, the efficiency restoring policy scheme described in Proposition 4.10 is the same as that in durable good markets with commitment ability of producers presented in row 5 of Table 4.1 in the vintage model of Sect. 4.1: In every period, the disposal externality is internalized by a Pigouvian waste tax whereas the stock subsidy corrects the market power of durable good producers.

Things are more complicated in case of imperfect sales markets without commitment ability ($P' < 0$, $\varphi = 1$). For this scenario, a regulatory scheme satisfying (4.69)–(4.71) is specified in

Proposition 4.11. *Under imperfect competition, the sales equilibrium is efficient if*

$$\tau_{w_1} = E'[n(1 - \phi^o)c_1^o] > 0, \qquad \tau_{w_2} = E'(nc_2^o) > 0,$$

$$\tau_\phi = \rho c_1^o z \begin{cases} < 0 & if \quad n = 1, \\ \gtreqless 0 & if \quad n > 1, \end{cases}$$

$$\tau_{c_1} = c_1^o P'(nc_1^o) + \rho\phi^o z \begin{cases} < 0 & if \quad n = 1, \\ \gtreqless 0 & if \quad n > 1, \end{cases}$$

$$\tau_{c_2} = c_2^o P'(n c_2^o) < 0 \,,$$

$$\tau_{y_t} = 0 \quad for \quad t \in \{1, 2\} \,.$$

In the absence of commitment ability in imperfectly competitive sales markets, the regulator still tackles the disposal externality with a Pigouvian waste tax in both periods. But in view of the remaining instruments, there are two important differences compared to the scheme described in Proposition 4.10 which represents the efficient scheme in rental markets without commitment ability (or, equivalently, in rental and sales markets with commitment ability).

First, the tax rate on product durability in the scheme for sales markets is not equal to zero, in general. In selling monopoly, the term z in the equilibrium condition (4.67) is unambiguously negative due to the incentive for planned obsolescence (distortion D4). Product durability is then inefficiently small even if Pigouvian waste taxes completely internalize the disposal externality. Thus, the regulator has to subsidize durability ($\tau_\phi < 0$) to induce the selling monopolist to render her products more durable. In selling oligopoly, the same argument holds if the incentive for planned obsolescence (distortion D4) dominates the incentive for using durability to absorb future willingness-to-pay of consumers (distortion D5). In this case, z is negative and durability has still to be subsidized. However, if the absorbing incentive D5 dominates the obsolescence incentive D4, then z is positive and the regulator employs a durability tax $\tau_\phi > 0$ in order to reduce inefficiently high durability.

Second, not only the tax rate on durability is adjusted by a term containing z, but also the tax rate on the first-period stock of the durable good. In selling monopoly, the regulator continues subsidizing the first-period stock ($\tau_{c_1} < 0$) since z in τ_{c_1} solely reflects the incentive for promoting future sales by decreasing first-period output (distortion D2) and, hence, z is unambiguously negative. But in selling oligopoly, the stock subsidy τ_{c_1} may be turned into a stock tax since z in τ_{c_1} also reflects distortion D3 which represents the firm's incentive to absorb future willingness-to-pay by raising first-period output. Consequently, if waste taxes equal marginal disposal costs and if z is positive and large enough to outweigh the market power distortion D1 represented by the term $c_1^o P'(n c_1^o)$, then the first-period stock of the durable good tends to be inefficiently large and has to be taxed in order to take on its efficient value.

The derivation of the efficient regulatory scheme in Proposition 4.11 presupposes the regulator is able to observe product durability. As already argued in Sect. 3.3, however, measuring and monitoring durability in real world markets is far more complicated as suggested by our formal model. Consequently, policy makers often do not make use of durability taxes or subsidies. This observation raises the question whether in imperfectly competitive sales markets without commitment ability of producers the welfare optimum can be attained without a tax or subsidy on durability. The answer is given in

Proposition 4.12. *Under imperfect competition, the sales equilibrium is efficient if*

$$\tau_{w_1} = E'[n(1 - \phi^o)c_1^o] - \rho z \begin{cases} > 0 & if \quad n = 1\,, \\ \gtreqless 0 & if \quad n > 1\,, \end{cases}$$

$$\tau_{w_2} = E'(nc_2^o) > 0\,,$$

$$\tau_{c_1} = c_1^o P'(nc_1^o) + \rho z \begin{cases} < 0 & if \quad n = 1\,, \\ \gtreqless 0 & if \quad n > 1\,, \end{cases}$$

$$\tau_{c_2} = c_2^o P'(nc_2^o) < 0\,,$$

$$\tau_\phi = \tau_{y_t} = 0 \qquad for \quad t \in \{1, 2\}\,.$$

According to Proposition 4.12, the regulator is able to achieve the efficient allocation in sales markets without commitment ability even if durability regulation is not viable. The difference to the regulatory scheme described in Proposition 4.11 is that the distortions D4 and D5 are no longer taken care of by the durability tax, but by the first-period tax rate on solid waste. Owing to this change in tax rates, not only the sign of the stock tax in period 1 is indeterminate, but also is the sign of the first-period waste tax: In selling monopoly, z is negative such that first-period waste still needs to be taxed ($\tau_{w_1} > 0$) and the first-period stock of the durable good still needs to be subsidized ($\tau_{c_1} < 0$). But in selling oligopoly, the tax rates on solid waste and on the durable's stock may be of either sign depending on the relative size of z in relation to marginal disposal costs $E'(\cdot)$ and in relation to the firm's market power reflected by the term $c_1^o P'(\cdot)$. If z is positive, absolutely greater than the market imperfection $c_1^o P'(\cdot)$ and smaller than marginal disposal costs $E'(\cdot)$, then the regulator restores efficiency in period 1 by a tax-tax scheme, i.e. a tax on first-period solid waste and a tax on the first-period stock of the durable good. This scheme is of special interest from a political point of view: As already mentioned in the introduction, a large part of the economic literature on environmental policy under imperfect competition is concerned with second-best emissions taxation since a subsidy on the output or the stock of the durable good is often considered politically infeasible. In contrast, our analysis shows that under special circumstances the regulator implements the first-best allocation with a tax on solid waste and a tax (rather than a subsidy) on the stock of the durable good. Admittedly, this result only holds under special circumstances. However, it suggests that first-best considerations are not superfluous per se. The regulator should first check the conditions for a tax-tax-result. She is then in the position to decide whether the welfare optimum can be implemented by waste and stock taxation or whether she ought to settle for the second-best outcome achieved by suitable emissions taxes.

To sum up the discussion in this section, the lack of commitment ability has no implications for efficient regulation under perfect competition and for regulation in rental markets. But in case of imperfectly competitive sales markets without commitment ability, the disposal externality and the market power of firms are not the only distortions and, therefore, additional regulation has to deal with the other distortions.

4.2.4 Second-Best Taxation

This section conducts a second-best analysis for the case that the regulator does not dispose of all instruments needed to restore efficiency under imperfect competition. For simplicity, we focus on second-best waste taxation and suppose $\tau_{y_t} = \tau_{c_t} = \tau_\phi = 0$ for $t \in \{1,2\}$ throughout. In Sect. 4.1, we found that the second-best waste tax in a durable good economy with commitment ability is smaller than marginal disposal costs provided the equilibrium stock of the durable good is a decreasing function of the waste tax (which seems to be plausible). The main objective of the present section is to check whether this underinternalization result carries over to the case without commitment ability of producers. In doing so, the rental and sales cases are treated separately.

Rental Markets

For the purpose of deriving the second-best waste taxes, the social planner makes use of the information that the durable stocks and product durability are functions of the waste taxes $\tau := (\tau_{w_1}, \tau_{w_2})$. In the rental case, these functions are determined by the equilibrium conditions (4.54)–(4.56) and denoted by $\phi^r(\tau)$, $c_1^r(\tau)$ and $c_2^r(\tau)$. The social planner maximizes social welfare with respect to τ_{w_1} and τ_{w_2}. Social welfare equals consumer benefits less production and disposal costs in both periods. Hence, in the symmetric rental equilibrium, it reads

$$
V(\tau) := S\Big[nc_1^r(\tau)\Big] - \kappa\Big[\phi^r(\tau)\Big]nc_1^r(\tau) - E\Big[n\Big[1 - \phi^r(\tau)\Big]c_1^r(\tau)\Big]
$$

$$
+ \rho\Big[S\Big[nc_2^r(\tau)\Big] - \bar{\kappa}n\Big[c_2^r(\tau) - \phi^r(\tau)c_1^r(\tau)\Big] - E\Big[nc_2^r(\tau)\Big]\Big] . \quad (4.72)
$$

Maximizing V with respect to τ_{w_1} and τ_{w_2} yields FOCs which with the help of the equilibrium conditions (4.54)–(4.56) simplify to

$$
\rho\frac{\partial c_2^r}{\partial \tau_{w_t}}\Big[\tau_{w_2} - E_2' - c_2^r P_2'\Big] - c_1^r\frac{\partial \phi^r}{\partial \tau_{w_t}}\Big[\tau_{w_1} - E_1'\Big]
$$

$$
+ \frac{\partial c_1^r}{\partial \tau_{w_t}}\Big[(1 - \phi^r)\tau_{w_1} - (1 - \phi^r)E_1' - c_1^r P_1'\Big] = 0 , \quad t \in \{1,2\} . \quad (4.73)
$$

From the equilibrium conditions (4.54)–(4.56) we obtain

$$\frac{\partial c_1^r}{\partial \tau_{w_1}} = \frac{1 - \phi^r}{(n+1)P_1' + nc_1^r P_1''} < 0 \,, \quad \frac{\partial c_2^r}{\partial \tau_{w_1}} = 0 \,, \quad \frac{\partial \phi^r}{\partial \tau_{w_1}} = \frac{1}{\kappa''} > 0 \,, \quad (4.74)$$

$$\frac{\partial c_1^r}{\partial \tau_{w_2}} = 0 \,, \quad \frac{\partial c_2^r}{\partial \tau_{w_2}} = \frac{1}{(n+1)P_2' + nc_2^r P_2''} < 0 \,, \quad \frac{\partial \phi^r}{\partial \tau_{w_2}} = 0 \,, \quad (4.75)$$

where the signs of the derivatives are due to assumption A4 and $\kappa'' > 0$. Note that the equilibrium stock in period t unambiguously decreases in the waste tax rate of period t since we suppose the cost function to be linear in output (unlike in the vintage model presented in Sect. 4.1). Using in (4.73) those derivatives from (4.74) and (4.75) which are equal to zero and solving the resulting expressions with respect to τ_{w_1} and τ_{w_2} yields the second-best waste taxes

$$\tau_{w_1}^{sb} = E_1' + c_1^r P_1' \frac{\partial c_1^r / \partial \tau_{w_1}}{\partial w_1^r / \partial \tau_{w_1}} \quad \text{and} \quad \tau_{w_2}^{sb} = E_2' + c_2^r P_2' \,. \quad (4.76)$$

In period 1, we obtain the same expression for the second-best waste tax as in the model with commitment ability in Sect. 4.1. This can be seen by comparing $\tau_{w_1}^{sb}$ from (4.76) with τ_w^{sb} from p. 105. (4.74) and (4.75) imply $\partial w_1^r / \partial \tau_{w_1} = (1 - \phi^r)(\partial c_1^r / \partial \tau_{w_1}) - c_1^r(\partial \phi_1^r / \partial \tau_{w_1}) < 0$ and, thus, (4.76) proves

Proposition 4.13. *Suppose the durable is rented under conditions of imperfect competition. Then $\tau_{w_t}^{sb} < E_t'$ for $t \in \{1, 2\}$.*

Underinternalization is second-best optimal in rental markets without precommitment. The reason is that distortions D2–D5 are absent in the rental case. Hence, the rental equilibrium without precommitment is distorted by the same two imperfections as in the presence of commitment ability, namely the disposal costs externality and distortion D1 which reflects market power of firms. The intuition of the underinternalization result in Proposition 4.13 is therefore also the same as in the case with precommitment: If waste taxes in both periods are set equal to marginal disposal costs, then the disposal costs externality is completely internalized and product durability is efficient due to Swan's independence result whereas the stock of the durable good and, hence, also waste in both periods are inefficiently small owing to the market power distortion D1. If waste taxes in both periods are increased beyond marginal disposal costs, then social welfare declines since durability increases beyond its efficient level according to $\partial \phi^r / \partial \tau_{w_1} > 0$ in (4.74) and since the stock of the durable good further decreases below its efficient level in both periods according to $\partial c_t^r / \partial \tau_{w_t} < 0$ in (4.74) and (4.75). Hence, waste taxes greater than marginal disposal costs cannot be second-best optimal. Following the usual second-best argument, however, society gains from setting waste taxes below marginal disposal costs. Durability then becomes inefficiently small, indeed, but at the same time, the stock of the durable good becomes greater and more efficient.

Sales Markets

Now turn to sales markets without commitment ability of durable good producers. In this case, social welfare may be written as in (4.72) except that $\phi^r(\tau)$, $c_1^r(\tau)$ and $c_2^r(\tau)$ are replaced by $\phi^s(\tau)$, $c_1^s(\tau)$ and $c_2^s(\tau)$ which represent the equilibrium values in sales markets as functions of the waste tax rates. These functions are determined by the equilibrium conditions (4.62), (4.66) and (4.67). Differentiating social welfare with respect to the waste tax rates and using the equilibrium conditions to cancel common terms yields the FOCs

$$\rho \frac{\partial c_2^s}{\partial \tau_{w_t}} \left[\tau_{w_2} - E_2' - (c_2^s - \phi^s c_1^s) P_2' \right] - c_1^s \frac{\partial \phi^s}{\partial \tau_{w_t}} \left[\tau_{w_1} - E_1' + \rho z \right]$$

$$+ \frac{\partial c_1^s}{\partial \tau_{w_t}} \left[(1 - \phi^s) \tau_{w_1} - (1 - \phi^s) E_1' - c_1^s P_1' - \rho \phi^s z \right] = 0, \quad t \in \{1, 2\} . \quad (4.77)$$

Defining $\eta_t^x := (\partial x^s / \partial \tau_{w_t}) \tau_{w_t} / x^s$ as the elasticity of the variable $x^s \in \{c_1^s, c_2^s, \phi^s, w_1^s\}$ with respect to τ_{w_t} and solving (4.77) with respect to τ_{w_1} and τ_{w_2}, we obtain the second-best waste taxes

$$\tau_{w_1}^{sb} = E_1' + c_1^s P_1' \frac{\eta_1^{c_1} \eta_2^{c_2} - \eta_2^{c_1} \eta_1^{c_2}}{(1 - \phi^s)(\eta_1^{w_1} \eta_2^{c_2} - \eta_2^{w_1} \eta_1^{c_2})}$$

$$+ \rho \phi^s z \frac{(\eta_1^{c_1} + \eta_1^{\phi}) \eta_2^{c_2} - (\eta_2^{c_1} + \eta_2^{\phi}) \eta_1^{c_2}}{(1 - \phi^s)(\eta_1^{w_1} \eta_2^{c_2} - \eta_2^{w_1} \eta_1^{c_2})} , \quad (4.78)$$

$$\tau_{w_2}^{sb} = E_2' + (c_2^s - \phi^s c_1^s) P_2' + c_1^s P_1' \frac{c_1 (\eta_2^{c_1} \eta_1^{w_1} - \eta_1^{c_1} \eta_2^{w_1})}{\rho c_2^s (\eta_2^{c_2} \eta_1^{w_1} - \eta_1^{c_2} \eta_2^{w_1})}$$

$$+ \rho \phi^s z \frac{c_1^s [(\eta_2^{c_1} + \eta_2^{\phi}) \eta_1^{w_1} - (\eta_1^{c_1} + \eta_1^{\phi}) \eta_2^{w_1}]}{\rho c_2^s (\eta_2^{c_2} \eta_1^{w_1} - \eta_1^{c_2} \eta_2^{w_1})} . \quad (4.79)$$

According to (4.78) and (4.79), the second-best waste tax in each period equals marginal disposal costs in that period plus some additional terms which are indeterminate in sign, in general. One may be inclined to conclude on the basis of this insight that overinternalization can be second-best optimal in sales markets without commitment ability. Such a conclusion would be somewhat unsatisfactory since it does not provide information about the reasons for the second-best tax being greater than marginal disposal costs. But considering a sequence of special cases will reveal the mechanisms which lead to overinternalization.

Consider the special case of a durable good monopoly ($n = 1$) with exogenously given product durability (ϕ fixed) as a benchmark. In a slightly different framework, this case is already analyzed in the article of Goering and Boyce

(1997) which we referred to in the introduction already on p. 10.[14] According to their analysis, a necessary condition for the second-best tax to exceed marginal damage is that the cost function is strictly concave in output. In the present durable good model, this necessary condition for overinternalization is violated since we assume linear production costs. Hence, it is expected that the optimal waste taxes (4.78) and (4.79) for $n = 1$ and given ϕ fall short of marginal disposal costs. Formally, this is checked by observing that fixed durability implies $\eta_t^\phi = 0$ and $\eta_t^{w_1} = \eta_t^{c_1}$ for $t \in \{1, 2\}$. The second-best waste taxes (4.78) and (4.79) then become

$$\tau_{w_1}^{sb} = E_1' + \frac{c_1^s P_1' + \rho \phi^s z}{1 - \phi^s} \quad \text{and} \quad \tau_{w_2}^{sb} = E_2' + (c_2^s - \phi^s c_1^s) P_2' \,. \quad (4.80)$$

$\tau_{w_2}^{sb}$ is smaller than E_2' since we assume an interior solution $y_2^s = c_2^s - \phi^s c_1^s > 0$. Furthermore, case 4 in Table 4.2 shows that in the fixed durability monopoly, we have $z < 0$. This together with (4.80) proves

Proposition 4.14. *Suppose the durable is sold under conditions of imperfect competition, ϕ is exogenously given and $n = 1$. Then $\tau_{w_t}^{sb} < E_t'$ for $t \in \{1, 2\}$.*

In line with the result derived by Goering and Boyce (1997), the second-best waste taxes in the durable good monopoly with fixed durability and without precommitment fall short of marginal disposal costs in both periods. The intuition becomes obvious by considering case 4 in Table 4.2: Under the conditions of Proposition 4.14, the distortions D4 and D5 are not present in the durable good market since product durability is fixed. Moreover, the incentive D3 for revenue raising is absent since it is effective only in oligopoly, but not in monopoly. The distortions which remain are D1 and D2. These distortions tend to shift first-period and second-period output below the efficient levels. If in such a scenario the waste taxes in both periods are set equal to marginal disposal costs, then the disposal costs externality is completely internalized, indeed, but output is inefficiently low in both periods owing to the market imperfections D1 and D2. Tightening environmental policy would harm social welfare since firms would further cut down their activity levels. Thus, overinternalization cannot be second-best in the fixed durability monopoly. According to the usual second-best argument, however, society gains from fixing the waste taxes below marginal disposal costs since output then becomes greater and more efficient.

In view of the benchmark case considered in Proposition 4.14, we are able to identify the conditions under which the underinternalization result is reversed. First, relax the assumption of monopoly and consider a durable good oligopoly ($n > 1$) with exogenously given durability (ϕ fixed). For this set of assumptions, the optimal waste taxes are again captured by (4.80) since fixed durability implies $\eta_t^\phi = 0$ and $\eta_t^{w_1} = \eta_t^{c_1}$ for $t \in \{1, 2\}$. But in contrast to the

---

[14] Goering and Boyce (1997) consider production emissions instead of consumption waste and exclusively focus on the monopoly case with fixed durability.

benchmark case, case 2 in Table 4.2 shows that z is not necessarily negative. Consequently, $\tau_{w_1}^{sb}$ in (4.80) exceeds marginal disposal costs if it can be shown that z is positive and larger than $-c_1^s P_1'/\rho\phi^s$. Such a constellation is contained in

Proposition 4.15. *Suppose the durable is sold under conditions of imperfect competition, ϕ is exogenously given and $n > 1$. Then $\tau_{w_2}^{sb} < E_2'$. But there are model specifications for which $\eta_t^{w_t}, \eta_t^{c_t}, \eta_t^{y_t} < 0$, $t \in \{1,2\}$, and $\tau_{w_1}^{sb} > E_1'$.*

Proof. For the durable good oligopoly with fixed durability, the market equilibrium condition (4.67) is dropped. So, the pertinent equilibrium conditions are (4.62) and (4.66). $\tau_{w_2}^{sb} < E_2'$ immediately follows from (4.80). In order to prove the remainder of Proposition 4.15, it is sufficient to provide an example. Suppose a linear specification of our model, i.e. $P(C_t) = \beta_1 - \beta_2 C_t$ and $E(W_t) = \gamma W_t$ with $\beta_1, \beta_2, \gamma > 0$. Fixed durability implies fixed unit costs $\kappa > 0$ and $\bar{\kappa} > 0$. Solving the equilibrium conditions (4.62) and (4.66) for the equilibrium stocks c_1^s and c_2^s as functions of the waste taxes τ_{w_1} and τ_{w_2} and computing the partial derivatives yields

$$\frac{\partial c_1^s}{\partial \tau_{w_1}} = -\frac{(1-\phi)(n+1)^2}{\beta_2[\rho\phi^2(n^2+1)+(n+1)^3]} < 0, \tag{4.81}$$

$$\frac{\partial c_1^s}{\partial \tau_{w_2}} = -\frac{\rho\phi(n-1)}{\beta_2[\rho\phi^2(n^2+1)+(n+1)^3]} < 0, \tag{4.82}$$

$$\frac{\partial c_2^s}{\partial \tau_{w_1}} = -\frac{\phi(1-\phi)(n+1)}{\beta_2[\rho\phi^2(n^2+1)+(n+1)^3]} < 0, \tag{4.83}$$

$$\frac{\partial c_2^s}{\partial \tau_{w_2}} = -\frac{(n+1)^2+\rho\phi^2 n}{\beta_2[\rho\phi^2(n^2+1)+(n+1)^3]} < 0. \tag{4.84}$$

Equation (4.81)–(4.84) imply

$$\frac{\partial w_1^s}{\partial \tau_{w_1}} = (1-\phi)\frac{\partial c_1^s}{\partial \tau_{w_1}} < 0, \quad \frac{\partial w_2^s}{\partial \tau_{w_2}} = \frac{\partial c_2^s}{\partial \tau_{w_2}} < 0, \quad \frac{\partial y_1^s}{\partial \tau_{w_1}} = \frac{\partial c_1^s}{\partial \tau_{w_1}} < 0, \tag{4.85}$$

and

$$\frac{\partial y_2^s}{\partial \tau_{w_2}} = \frac{\partial c_2^s}{\partial \tau_{w_2}} - \phi^s\frac{\partial c_1^s}{\partial \tau_{w_2}} = -\frac{(n+1)^2+\rho\phi^2}{\beta_2[\rho\phi^2(n^2+1)+(n+1)^3]} < 0. \tag{4.86}$$

Owing to the signs in (4.81)–(4.86), we obtain $\eta_t^{w_t}, \eta_t^{c_t}, \eta_t^{y_t} < 0$ for $t \in \{1,2\}$. It remains to compute the second-best waste taxes. For $n > 1$ and fixed ϕ, they are captured by (4.80). $\tau_{w_2}^{sb}$ therefore always falls short of marginal disposal costs (independent of the linear model specification). But inserting the equilibrium c_1^s and c_2^s for the linear specification into (4.80) reveals that $\tau_{w_1}^{sb}$ may exceed marginal disposal costs. For example, if $\beta_1 = 20$, $\beta_2 = 0.2$, $\gamma = \bar{\kappa} = 1$, $\kappa = 15$, $\rho = 0.97$, $\phi = 0.6$ and $n = 20$, then $c_1^s \approx 1.296$, $c_2^s \approx 4.500$, $\tau_{w_2}^{sb} \approx 0.256 < 1 = \gamma = E_2'$ and $\tau_{w_1}^{sb} \approx 1.322 > 1 = \gamma = E_1'$.   □

Proposition 4.15 shows that in the durable good oligopoly with exogenous durability and without commitment ability, overinternalization in the first-period may be second-best optimal. The intuition of such an overinternalization result becomes evident from case 2 of Table 4.2: As in the case of fixed durability monopoly considered in Proposition 4.14, the distortions D4 and D5 vanish for the durable goods oligopoly with given durability. But the incentive for revenue raising (distortion D3) is an essential feature of the sales equilibrium under oligopolistic market structure. This effect also diverts first-period output from its efficient level either, but points into the opposite direction as D1 and D2. Eliminating the disposal externality through setting waste taxes equal to the marginal disposal costs leads again to an inefficiently low second-period output, but the first-period output is inefficiently large if the impact of D1 and D2 is overcompensated by the impact of revenue raising D3. Welfare is higher for a second-period tax that is fixed below marginal disposal costs and a first-period tax set *above* marginal disposal costs since second-period output increases ($\eta_2^{y_2} < 0$) and first-period output declines ($\eta_1^{y_1} < 0$) so that both variables become more efficient. Therefore, overinternalization is second-best in period 1.

Another overinternalization result is obtained when we return to monopoly ($n = 1$), but now suppose that durability is an endogenous choice variable of producers (ϕ endogenous). In this case, we obtain

Proposition 4.16. *Suppose the durable is sold under conditions of imperfect competition, ϕ is endogenous and $n = 1$. Then there exist model specifications for which $\eta_t^{w_t}, \eta_t^{c_t}, \eta_t^{y_t} < 0$, $t \in \{1, 2\}$, $\eta_1^{\phi} > 0$, $\tau_{w2}^{sb} < E_2'$ and $\tau_{w1}^{sb} > E_1'$.*

Proof. For proving Proposition 4.16, it is again sufficient to present an example. We now suggest a linear-quadratic specification of the model. As in the proof of Proposition 4.15, demand and damage functions are assumed to be linear. Furthermore, we suppose first-period unit costs are quadratic, i.e. $\kappa(\phi) = \omega\phi^2$ with $\kappa'' = 2\omega > 0$. The market equilibrium conditions for the durable good monopoly with endogenous durability are (4.62), (4.66) and (4.67) with $z < 0$. These conditions determine the equilibrium stocks of the durable good and the equilibrium durability as functions of the waste tax rates. The partial derivatives are computed as

$$\frac{\partial c_1^s}{\partial \tau_{w1}} = -\frac{4\omega\beta_2(1 - \phi^s) + \rho c_1^s \beta_2^2}{|J^s|} < 0 , \qquad \frac{\partial \phi^s}{\partial \tau_{w1}} = \frac{4\beta_2^2 + \rho\phi^s\beta_2^2}{|J^s|} > 0 , \quad (4.87)$$

$$\frac{\partial c_2}{\partial \tau_{w2}} = -\frac{4\omega\beta_2 + \rho c_1^s \beta_2^2 + \omega\beta_2\rho\phi^{s2}}{|J^s|} < 0, \qquad \frac{\partial c_1^s}{\partial \tau_{w2}} = \frac{\partial \phi^s}{\partial \tau_{w2}} = 0 , \qquad (4.88)$$

with the Jacobian determinant $|J^s| := 8\omega\beta_2^2 + 2\rho c_1^s \beta_2^3 + 2\omega\beta_2^2\rho\phi^{s2} > 0$. Equations (4.87) and (4.88) imply

$$\frac{\partial w_1^s}{\partial \tau_{w1}} = (1 - \phi^s)\frac{\partial c_1^s}{\partial \tau_{w1}} - c_1^s \frac{\partial \phi^s}{\partial \tau_{w1}} < 0 , \qquad \frac{\partial w_2^s}{\partial \tau_{w2}} = \frac{\partial c_2^s}{\partial \tau_{w2}} < 0 ,$$

$$\frac{\partial y_1^s}{\partial \tau_{w_1}} = \frac{\partial c_1^s}{\partial \tau_{w_1}} < 0, \qquad \frac{\partial y_2^s}{\partial \tau_{w_2}} = \frac{\partial c_2^s}{\partial \tau_{w_2}} < 0.$$

Hence, for the linear-quadratic specification, we obtain $\eta_1^\phi > 0$ and $\eta_t^{w_t}, \eta_t^{c_t}, \eta_t^{y_t} < 0$ for $t \in \{1, 2\}$. It remains to determine the second-best waste taxes. From (4.88) follows $\eta_2^{c_1} = \eta_2^\phi = 0$. Inserting this result and the parametric functions into (4.78) and (4.79) yields

$$\tau_{w_1}^{sb} = \gamma - \beta_2 c_1^s \frac{2\omega\beta_2(1 - \phi^s)(2 + \rho\phi^{s2}) - \rho c_1^s \beta_2^2 (2\phi^s - 1)}{4\omega\beta_2(1 - \phi^s)^2 + 4c_1^s\beta_2^2 + \rho c_1^s \beta_2^2}, \qquad (4.89)$$

$$\tau_{w_2}^{sb} = \gamma - \beta_2(c_2^s - \phi^s c_1^s). \qquad (4.90)$$

The second-best waste tax in period 2 falls short of marginal disposal costs γ due to the assumption of an interior equilibrium $y_2^s = c_2^s - \phi^s c_1^s > 0$. However, the difference between the second-best tax and marginal disposal costs in period 1 is ambiguous in sign since $\phi^s \gtreqless 1/2$. To provide an example for overinternalization, we compute (4.89) and (4.90) for $n = 1$, $\rho = 0.99$, $\beta_1 = 10$, $\beta_2 = \omega = \bar{\kappa} = 1$ and $\gamma = 2$. We then obtain $c_1^s \approx 4.171$, $c_2^s \approx 7.000$, $\phi^s \approx 0.815$, $\tau_{w_2}^{sb} \approx -1.602 < 2 = \gamma = E_2'$ and $\tau_{w_1}^{sb} \approx 2.321 > 2 = \gamma = E_1'$. $\quad\square$

For the overinternalization result derived in Proposition 4.16, the firm's incentive for planned obsolescence is of crucial importance: As shown in case 3 of Table 4.2, the incentives D3 and D5 for revenue raising are not present in the durable good monopoly with endogenous durability since the monopolist does not face competition. The remaining distortions are D1, D2 and D4. D1 and D2 give rise to insufficient production rates in both periods while the incentive for planned obsolescence (distortion D4) tends to induce an underprovision of durability. Consequently, for waste taxes set equal to marginal disposal costs, not only output in both periods is inefficiently small, but also is product durability. An increase in the first-period waste taxes beyond marginal disposal costs does not necessarily reduce social welfare. The negative effect of further decreasing the production rates ($\eta_t^{y_t} < 0$ for $t \in \{1, 2\}$) is accompanied by the positive welfare effect of increasing durability towards its efficient level ($\eta_1^\phi > 0$) since the firms reduce waste tax payments by rendering their products more durable. If the latter effect overcompensates the former effect, then the optimal first-period tax rate is greater than marginal disposal costs.

To sum up the discussion in this section, the lack of commitment ability may lead to overinternalization provided the durable goods are sold by their producers. Observe that there is a crucial difference between the overinternalization results in Propositions 4.15 and 4.16 and the overinternalization results derived in the articles referred to in the introduction. In these articles, a *necessary* condition for the second-best emissions tax to exceed marginal damage is either (a) that production costs are strictly concave in output (Goering and Boyce, 1997) or (b) that the firm's equilibrium output (stock of the durable good) or the firm's equilibrium emissions rise with increasing emissions tax rates (Ebert, 1992; Simpson, 1995; Katsoulacos and Xepapadeas, 1995; Requate, 1997, and the second-best results presented in equation (3.76) and Proposition 4.5 of the present study). Empirically, production technologies are likely to be characterized by constant or slightly diminishing returns to scale with the consequence that the cost function is linear or convex in output (Baily et al., 1992; Burnside, 1996). Furthermore, some studies indicate that firms restrict their production and emissions when regulated by environmental policy (Brännlund and Gren, 1999). In sum, there is some evidence that the necessary conditions (a) and (b) for overinternalization are not satisfied (at least in some real world markets). However, the overinternalization cases identified in Proposition 4.15 and 4.16 require neither a concave cost function (since throughout the analysis we assume the cost function to be linear in output) nor an abnormal correlation between the waste tax and the firm's output, stock or waste of the durable good (since $\eta_t^{w_t}, \eta_t^{c_t}, \eta_t^{y_t}$ are all negative).

4.3 Implications for Practical Waste Policy

This section offers a number of different interpretations of the theoretical model in order to evaluate various policy instruments for solid waste management.[15] As already announced in the introduction to the present chapter, we will distinguish between instruments which are in accordance with the principle of extended producers responsibility (EPR instruments) and instruments which intervene at the level of households or disposal firms and thus do not follow the EPR principle (non-EPR instruments). In evaluating these instruments, it is important to keep in mind the following two features of real world disposal markets. First, total disposal costs E can be split up into private disposal costs E_p (costs of land, capital, labor and energy inputs of disposal facilities) and social disposal costs E_s (environmental costs due to the emissions of pollutants from disposal facilities like e.g. emissions of an incinerator or leakage of a landfill). Second, disposal facilities are either under public or private control. In calculating disposal prices, unregulated private disposal firms only take into account private marginal disposal costs E_p' whereas public disposal firms are typically required to include total marginal disposal costs $E' = E_p' + E_s'$ in their calculation.

We will begin our discussion of environmental policy with non-EPR instruments under perfect competition, then turn to EPR instruments under perfect competition and finally evaluate EPR under imperfect competition. As implied by the formal models of Sects. 4.1 and 4.2, under perfect competition, regulation of rental markets is the same as regulation of sales markets independent of whether firms possess commitment ability or not. For simplicity,

---

[15] This section borrows heavily from Runkel (2003b).

we therefore do not distinguish between rental markets and sales markets and refer only to the results derived in the model with precommitment in Sect. 4.1 although qualitatively the same results are obtained in the model without commitment ability in Sect. 4.2. Under imperfect competition, in contrast, regulation of rental markets and sales markets is only the same if producers precommit. Hence, in case of imperfect competition, we refer to the results of the models in both Sect. 4.1 and Sect. 4.2.

Non-EPR Instruments Under Perfect Competition

We start by focusing on some non-EPR instruments and the case of perfect competition. The non-EPR instruments are summarized in Table 4.3. In many

Table 4.3. Policy Parameters under Different Non-EPR Instruments

row	non-EPR instrument	policy parameters
1	lump-sum fees (LSF)	$\tau_w = \tau_c = \tau_y = \tau_\phi = 0$
2	regulated private disposal markets (RDM)	$\tau_w = E'$, $\tau_c = \tau_y = \tau_\phi = 0$
3	unit-based pricing (UBP)	$\tau_w = E'$, $\tau_c = \tau_y = \tau_\phi = 0$
4	influencing demand behavior (IDB)	see rows 2 and 3 of Table 4.1

countries, the problem of disposal costs is tackled with *lump-sum fees* (LSF) levied on households and with public disposal firms whose costs are covered by the revenues from LSF (Golub, 1998). Since, by definition, flat fees are independent of the amount of waste the household generates, they do not provide incentives to reduce waste or to adjust the demand of the durable good. The model with LSF is therefore essentially equivalent to the model of the laissez faire economy ($\tau_w = \tau_c = \tau_y = \tau_\phi = 0$) which, according to Proposition 4.3i, is characterized by market failure with inefficiently low product durability. In the subsequent analysis, this market failure will serve as a benchmark for all other regulatory schemes.

One way to overcome market failure is to introduce *regulated private disposal markets* (RDM). Under this regime, the regulator replaces public disposal firms by private firms, levies taxes on environmental pollution of these firms at rates equal to marginal social costs E'_s and keeps households responsible for waste disposal. Disposal firms then account for the emissions taxes in calculating their disposal prices and, consequently, households have to pay disposal prices equal to total marginal disposal costs $E' = E'_p + E'_s$. By adjusting their willingness-to-pay for the durable good, rational utility-maximizing households pass on disposal costs to producers such that E' enters the producers' profit maximization problems although producers are not directly taxed under RDM. In our formal model, RDM can therefore be characterized by setting $\tau_w = E'$. According to Proposition 4.2 and row 1 of Table 4.1, this implies that RDM increase durability (compared to LSF) and implement the first-best optimum under perfect competition.

As pointed out by Staudt et al. (1997), RDM often do not work since substantial fixed costs render it unprofitable for private firms to enter disposal markets. As an alternative, the regulator may then correct market failure through *unit-based pricing* (UBP) as proposed by Jenkins (1993). Under this regime, every unit of waste generated by households is charged with its complete marginal disposal costs E', and the costs of disposal firms are covered by the revenues of UBP. Again, rational households pass on the unit-based fee to producers so that in terms of our formal model, UBP becomes also equivalent to $\tau_w = E'$. Consequently, UBP increases product durability according to Proposition 4.2 and restores the first-best optimum under perfect competition according to row 1 of Table 4.1.

Under both RDM and UBP, households are charged for their waste generation. According to Fullerton and Kinnaman (1995, 1996), charging households introduces incentives for moonlight dumping which destroy the efficiency properties of RDM and UBP. Another option for the regulator is then to directly *influence demand behavior* of households (IDB). For example, she may tax the quantity (stock) and subsidize durability of demanded products. By reducing the demand of the durable good and by demanding more durable products, rational households pass on the tax and subsidy to producers. Therefore, if the tax on the quantity (stock) and the subsidy on durability of the demanded goods are set as in row 2 of Table 4.1 (row 3 of Table 4.1), then producers are provided with the right incentives to raise durability and to implement the first-best optimum under perfect competition. The stock tax is of particular interest since in many countries stock taxes are already introduced, for example, the car tax in Germany which we already referred to in Sect. 3.3 on p. 84.

The problem of IDB is that it requires a subsidy on durability. In reality, durability is difficult to measure, in particular at the beginning of the product's life when the subsidy on durability is due. Moreover, Palmer and Walls (1999) argue that it is questionable whether households act rational. If they are boundedly rational, then the taxes and subsidies under RDM, UBP and IDB are not passed on to producers. In this case, producers do not receive price signals even if the regulator imposes the correct tax schemes on households and disposal firms. In our formal model, we then have to set all tax rates equal to zero so that we are back to the benchmark case with market failure and an inefficiently low durability. It is then preferable to regulate producers directly by EPR.

EPR Instruments Under Perfect Competition

In analyzing EPR instruments, we first investigate the case of perfect competition again and, thus, refer to the results in Sect. 4.1. Four different EPR instruments are considered and compared to the benchmark. These EPR instruments are characterized in Table 4.4. The benchmark is represented ei-

Table 4.4. Policy Parameters under Different EPR Instruments ($\tau_c = \tau_\phi = 0$)

row	EPR instrument	policy parameters
1	benchmark	$\tau_w = 0, \quad \tau_y = 0$
2	take-back requirement + public disposal firms (TBR)	$\tau_w = E', \quad \tau_y = 0$
3	take-back requirement + private disposal firms (TBR*)	$\tau_w = E'_p \in]0, E'[, \quad \tau_y = 0$
4	public waste fund (PWF)	$\tau_w = 0, \quad \tau_y = \Omega E'$
5	waste management organization (WMO)	$\tau_w = 0, \quad \tau_y = \Omega E'_p \in]0, \Omega E'[$

ther by LSF or by RDM, UBP and IDB with boundedly rational consumers so that all tax rates in our formal model are equal to zero. The first EPR instrument is a take-back requirement combined with setting up public disposal firms (TBR). According to Michaelis (1995), TBR mandates producers to take back worn-out products from consumers and to bear disposal costs. Since disposal firms are under public control, producers pay disposal prices equal to total marginal disposal costs E'. In our formal model, these disposal costs of producers are characterized by setting $\tau_w = E'$ and $\tau_y = 0$. The second EPR instrument in Table 4.4 is also a take-back requirement, but now combined with (unregulated) private disposal firms (TBR*). Since these firms only account for private disposal costs, producers have to pay disposal prices equal to E'_p. Hence, we represent TBR* in our formal model by $\tau_w = E'_p$ and $\tau_y = 0$. The third EPR instrument in Table 4.4 is a public waste fund (PWF) as proposed by Lidgren and Skogh (1996). Instead of paying disposal costs at the end of the products' life, the idea of PWF is that producers are charged with a unit-based fee upstream when the goods are produced and sold. The fee is paid to PWF and equals the present value of disposal costs caused by a unit of the durable good during its whole lifetime, $\Omega E'$. The disposal activities are then financed by the revenues of PWF. In terms of the formal model,

PWF is thus characterized by setting $\tau_w = 0$ and $\tau_y = \Omega E'$. The last EPR instrument in Table 4.4 is a private waste management organization (WMO) like the German Duales System Deutschland (Rousso and Shah, 1994). The functioning of WMO is akin to that of PWF: At the beginning of the products' life, WMO charges producers with a 'license fee' which is used to pay disposal costs at the end of the products' life. In contrast to PWF, however, WMO is privately organized and takes into account private disposal costs only. In our formal model, WMO is thus represented by $\tau_w = 0$ and $\tau_y = \Omega E_p'$.

In order to determine the effects of the EPR instruments on product durability under perfect competition, we compare Table 4.4 with the results of our theoretical model presented in Sects. 4.1 and 4.2. In the benchmark scenario, producers do not receive the right price signals and therefore generate an inefficient allocation with inefficiently low product durability according to Proposition 4.3i. In contrast, under the EPR instruments, the producers are charged (at least a part of) the disposal costs. They then take into account end-of-life costs of their durables and increase product durability compared to the benchmark as shown by Proposition 4.2. This is true for all EPR instruments listed in Table 4.4. The instruments differ only in the extent to which durability is increased. Under TBR, durability is greater than under TBR* since TBR forces producers to bear the total marginal disposal costs whereas TBR* assigns only private marginal disposal costs to producers. By an analogous argument, durability under PWF exceeds durability under WMO. Unfortunately, our analysis in Sect. 4.1 does not allow to compare durability achieved under TBR and TBR*, on the one hand, and under PWF and WMO, on the other hand, since this requires to numerically specify Ω and the impact of the tax rates on durability.

The effects of EPR on welfare under perfect competition are as follows. According to row 1 of Table 4.1 and row 2 of Table 4.4, the first-best optimum is restored with TBR since this instrument charges solid waste with complete marginal disposal costs E'. As implied by rows 1–4 of Table 4.1 and rows 3–5 of Table 4.4, the other EPR instruments are not capable of attaining the first-best outcome. Yet, these instruments also increase welfare in comparison to the benchmark: Under TBR*, τ_w is positive and smaller than marginal disposal costs E'. The second-best results derived in (4.39) and Proposition 4.5 show that such a tax increases welfare under perfect competition, provided the equilibrium stock of the durable good is decreasing in the waste tax rate. A similar argument applies to PWF and WMO since in these scenarios τ_y is positive and not greater than the present value $\Omega E'$ of marginal disposal costs. The consequence is that Proposition 4.6 and equation (4.40) imply a welfare increase under perfect competition. In sum, under perfect competition, EPR appears to be an appropriate measure either to reach the first-best optimum (with TBR) or at least to increase welfare compared to the benchmark (with TBR*, PWF or WMO).

EPR Instruments under Imperfect Competition

This positive appraisal of EPR becomes strained under imperfect competition. Let us start with the case of producers with commitment ability in which regulation of sales and rental markets is the same. According to Proposition 4.2, EPR increases product durability under imperfect competition, too, but the first-best is reached by none of the EPR instruments since the regulator employs one tax rate only whereas the first-best policy requires setting at least two tax rates different from zero according to rows 5–10 of Table 4.1. In comparison to the benchmark, EPR may even reduce welfare: Owing to Proposition 4.5, welfare under imperfect competition increases (decreases) as long as τ_w is smaller (greater) than a threshold value which, in turn, falls short of marginal disposal costs E' (if the equilibrium stock of the durable good is negatively correlated with the waste tax rate). Hence, at $\tau_w = E'$, the welfare function is downward sloping. This result combined with Table 4.4 implies that TBR decreases welfare under imperfect competition if marginal disposal costs E' are sufficiently large, i.e. if welfare at $\tau_w = E'$ is lower than at $\tau_w = 0$. Likewise, welfare under imperfect competition is reduced by TBR*, PWF and WMO if the terms E'_p, $\Omega E'$ and $\Omega E'_p$, respectively, are sufficiently large. The reason for the reduction in welfare is that EPR solely aims at internalizing disposal costs of solid waste while disregarding the market imperfection due to market power of firms.

Turning to durable good markets without commitment ability of producers, the above argumentation remains unchanged in case that firms rent their output since open-loop and closed-loop rental equilibria are identical. In contrast, for sales markets without precommitment, we derived the overinternalization results in Propositions 4.15 and 4.16. In such overinternalization cases, it is more likely that the EPR instruments do not bring about a welfare reduction, but a welfare improvement compared to the benchmark case since the welfare function attains its maximum at greater waste tax rates than in an underinternalization case. Anyway, whether the conditions for an overinternalization result are satisfied is ultimately an empirical question. To answer this question, we have to run the numerical computations with empirical values of the model parameters. Even though the economic literature provides estimates for price elasticities and particular costs parameters, unfortunately to date empirical studies are not available that inquire into the relation between product durability and production costs. Furthermore, quantifying the damage of industrial pollution reliably is still a difficult undertaking. Hence, realistic empirical estimates of the optimal emissions tax and the reliable comparative evaluation of EPR instruments is still on the agenda of further empirical research.

5

Recycling Policies for Consumer Durables

The previous chapter dealt with solid consumption waste and the associated disposal costs. In doing so, we have implicitly assumed that the disposal cost function also reflects the recycling of the durable consumption goods (e.g. on p. 89). Of course, such a modeling is rather abstract since it makes explicit neither the productivity of the recycled material in the production of the durable good nor the allocative consequences of product recyclability. This appears to be a severe simplification since there is some evidence that recycling is a waste management option of increasing importance. For example, the German agency for the environment estimates that in Germany 70-75% of the materials in wrecked cars is recycled (Umweltbundesamt, 1997, p. 442). The remaining 25-30% consist of plastics, glass and textiles for which no low cost recycling technology is yet available. But the German automobile manufactures signed a voluntary commitment to reduce these not yet recycled materials in their products to 15% in 2002 and 5% in 2015. In view of this evidence, it is necessary to develop and investigate models which explicitly account for aspects of durable good recycling.

For the case of nondurable goods, economic aspects of recyclability have received increasing attention in the economic literature e.g. by Fullerton and Wu (1998), Choe and Fraser (1999, 2001), Eichner and Pethig (2000, 2001) and Calcott and Walls (1999, 2000). An important insight of this literature is that an externality occurs when markets for recyclability are missing since producers then ignore the impact of recyclability on the recycling process. The most direct policy approach to internalize this externality is a Pigouvian subsidy on recyclability. The subsidy reflects marginal benefits of recyclability and provides the producers with the incentive for shifting recyclability up to its efficient level. Section 5.1 considers a vintage durable good model where, for simplicity, product durability is supposed to be fixed. We assume, however, that producers are able to vary product recyclability which influences the productivity of the recycling process. The main question to be addressed in this section is whether the policy recommendation of subsidizing recyclability carries over to the case of durable goods. We proceed by first characterizing the efficient allocation (Sect. 5.1.1), then identify a benchmark market economy in which all markets work frictionless (Sect. 5.1.2) and finally derive regulatory schemes capable of restoring efficiency in case the market for recyclability fails to be active (Sect. 5.1.3).

Section 5.2 extends the model of Sect. 5.1 in two directions. First, we introduce a functional relationship between durability and recyclability thus making durability an endogenous variable. As already argued in the introduction, such a relation is empirically relevant since both product attributes are related to the product's weight (Jambor and Beyer, 1997; Carle and Blount, 1999). Second, the waste-delaying effect introduced in Sect. 4.1 reveals that changes in durability inevitably raise problems of intergenerational distribution since less waste (less recycling input) today implies more waste (more recycling input) tomorrow. To address such problems, we consider two alternative types of time preferences in characterizing the efficient allocation: the Utilitarian and the Chichilnisky time preference. Under the Utilitarian approach, the social planner derives the efficient allocation by maximizing the present value of the consumers' utility where the present value is calculated with a constant and positive discount rate. The Utilitarian preference is based on an appealing set of axioms (Koopmans, 1960), but it discriminates against future generations due to the positive and constant discount rate. The Chichilnisky preference introduced by Chichilnisky (1996) satisfies two additional axioms which give more weight to the far distant future and, thus, treat the 'present' and the 'future' less unequally.

As in Sect. 5.1, in the extended model of Sect. 5.2, the main task also is to identify market failure and to propose policy recommendations in case that markets for product attributes are missing. We proceed in several steps again: First the efficient allocation under different kinds of time preferences is derived (Sect. 5.2.1), then we turn to a benchmark market economy where all markets work well (Sect. 5.2.2), and finally market failure and efficiency restoring policy options in case of absent markets for recyclability and durability are characterized (Sect. 5.2.3). Section 5.3 summarizes the main findings and discusses some practical aspects of the results.

5.1 Durable Goods Markets with Exogenous Product Durability

5.1.1 Model and Efficient Allocation

In contrast to the previous chapters, we restrict attention to the case of perfect competition and consider a general equilibrium model rather than a partial equilibrium model.[1] The model comprises production, consumption and recycling of a durable good with exogenously given product durability. To sharpen

---

[1] The analysis in this section is based on Eichner and Runkel (2001).

the focus on recycling, we simplify by ignoring throughout the environmental costs caused by solid waste of the durable good. In this subsection, the building blocks of the model are described in more detail and then the efficient allocation is characterized.

Production and Consumption

At time t, a representative producer manufactures a consumption good according to[2]

$$y(t) = Y\big[\,\underset{+}{\ell_y(t)}, \underset{+}{r(t)}, \underset{-}{\alpha(t)}\,\big] , \qquad (5.1)$$

where $Y : \mathbb{R}_+^2 \times [0,\bar{\alpha}] \mapsto \mathbb{R}_+$ is the twice continuously differentiable production function. The inputs in the production process are labor $\ell_y(t)$ and recycled material $r(t)$.[3] These inputs are used to produce $y(t)$ units of the consumption good with (built-in) recyclability $\alpha(t) \in [0,\bar{\alpha}]$ with $\bar{\alpha} > 0$. The product attribute recyclability can be interpreted in different ways. For example, suppose the total material input of the production process is a mix of different types of materials. $\alpha(t)$ may then be interpreted as homogeneity index of this aggregate material. For $\alpha(t) = \bar{\alpha}$, the aggregate material is homogeneous consisting of a single material. For $\alpha(t) = 0$, the aggregate material contains (a large variety of) different materials and is extremely heterogeneous. Products consisting of only one material are typically easier to recycle than products consisting of many different materials and, therefore, increasing the homogeneity index $\alpha(t)$ amounts to an increase in recyclability. An alternative interpretation is to think of $\alpha(t)$ as a disassembly index which captures information about the material interfaces. At $\alpha(t) = \bar{\alpha}$, the spent good can be very easily decomposed into its material components whereas at $\alpha(t) = 0$, the separation of materials is extremely difficult and, hence, recycling is extremely expensive. For either interpretation of recyclability, the firm needs to use up some inputs or needs to reduce output in order to make its products more recyclable. The derivative of the production function (5.1) with respect to recyclability $\alpha(t)$ is therefore negative.

The consumption good under consideration is assumed to be durable. As in Chaps. 2, 3 and Sect. 4.1, different vintages of the durable good are considered. Vintages differ in output and recyclability where $y(v)$ and $\alpha(v)$ are vintage v's output and durability, respectively. For simplicity, product durability is assumed to be fixed at the same level for all vintages. As a consequence, the decay function is $D(a)$ satisfying assumption A1. The total stock of the durable good at time t reads

---

[2] A plus or minus sign underneath an argument of a function denotes the sign of the respective partial derivative.

[3] As Fullerton and Wu (1998), in this section, we explicitly focus on recycled material and do not model virgin material. Virgin material is taken into account in the model of Sect. 5.2.

$$c(t) = \int_{-\infty}^{t} D(t - v)\, y(v) dv \, . \tag{5.2}$$

It equals the sum of the remaining units of all vintages $v \leq t$. The demand side is modeled by considering a representative consumer. As in the previous chapters, the consumer derives utility from the services of the durable, and the amount of services is proportional to the physical stock $c(t)$. Moreover, the consumer's utility is adversely affected by the endogenous labor supply $\ell(t)$. At time t, utility is therefore equal to

$$u(t) = U\big[\underset{+}{c(t)}, \underset{-}{\ell(t)}\big] \, , \tag{5.3}$$

where $U : \mathbb{R}_+^2 \mapsto \mathbb{R}_+$ is the twice continuously differentiable utility function of the representative consumer.

Recycling

After consumption, a fraction of the durable stock decays and turns into consumption residuals. For notational convenience, we denote the residuals of vintage v at time t by

$$w(v; t) = -D_a(t - v)y(v) \, . \tag{5.4}$$

Let $\mathcal{W}(t) := \{w(v; t) | \, v \leq t\}$ be the set of all consumption residuals at time t and let $\mathcal{A}(t) := \{\alpha(v) | v \leq t\}$ be the corresponding set of recyclability. Residuals are then recycled according to the recycling function

$$r(t) = \widetilde{R}\big[\underset{+}{\ell_r(t)}, \mathcal{W}(t), \mathcal{A}(t)\big] \, . \tag{5.5}$$

Equation (5.5) states that the amount of recycled material $r(t)$ depends on labor input $\ell_r(t)$, the amount of residuals $w(v; t)$ of all vintages $v \leq t$ and the recyclability $\alpha(v)$ of all vintages $v \leq t$. (5.5) is a rather general description of the recycling process since it implies that residuals and recyclability of every single vintage determine the recycling output.

  Though the realism of the generality of (5.5) is appealing, for reasons of tractability, the subsequent analysis focuses on an alternative modeling of the recycling technology. With respect to residuals, we assume that the total amount of residuals at time t

$$w(t) = \int_{-\infty}^{t} w(v; t) dv \tag{5.6}$$

enters the recycling function. As for recyclability, arguments of the recycling function are not the elements of the set $\mathcal{A}(t)$, but the (weighted) mean of the residuals' recyclability

$$\alpha_\mu(t) = \frac{\mu(t)}{w(t)} \qquad \text{with} \qquad \mu(t) = \int_{-\infty}^{t} w(v;t)\alpha(v)dv \;, \qquad (5.7)$$

and the (weighted) variance of the residuals' recyclability

$$\alpha_\sigma(t) = \frac{1}{w(t)} \int_{-\infty}^{t} w(v;t)\left[\alpha(v) - \alpha_\mu(t)\right]^2 dv \;,$$

which can conveniently be written as

$$\alpha_\sigma(t) = \frac{\sigma(t)}{w(t)} - \alpha_\mu^2(t) \qquad \text{with} \qquad \sigma(t) := \int_{-\infty}^{t} w(v;t)\alpha^2(v)dv \;. \qquad (5.8)$$

The recycling process is then described by

$$r(t) = R\big[\,\ell_r(t), w(t), \alpha_\mu(t), \alpha_\sigma(t)\,\big] \;, \qquad (5.9)$$
$${}_{+}\;\;\;\;{}_{+}\;\;\;\;\;{}_{+}\;\;\;\;{}_{+/-}$$

where $R : \mathbb{R}_+^2 \times [0, \bar\alpha]^2 \mapsto \mathbb{R}_+$ is the twice continuously differentiable recycling function. R in (5.9) is increasing in the total amount of residuals and in the mean of the residuals' recyclability. The sign of the derivative of R with respect to the variance of the residuals' recyclability is left unspecified. However, the main results of our analysis turn out to be independent of this sign. The advantage of the recycling function R from (5.9) over the recycling function $\widetilde{R}$ from (5.5) is that the recycling output is determined by some well-defined aggregate measures of residuals and their recyclability.

We offer two motivations for dealing with recycling functions of type (5.9). First, recyclers in real world markets are usually not able to observe the exact composition of residuals. Consider as an example a recycler who gets two truckloads of wrecked cars. The first contains cars constructed during 1965–1970 and the second contains cars constructed during 1970–1975. Based on her experience the recycler estimates the mean and variance of the recyclability of both car loads and based on this estimations she gives instructions to her employees. For example, if the recycler estimates mean recyclability of younger cars greater than that of older cars, then she assigns more working hours or more workers to the recycling of younger cars than to the recycling of older cars. Hence, the recycling output depends on the recycler's expectations about the residuals' recyclability expressed in terms of mean and variance of the residuals' recyclability. Hence, a recycling function of type (5.9) appears to be a realistic hypothesis.

The second (and perhaps more important) motivation for (5.9) is that, with some qualifications, the general recycling function $\widetilde{R}$ in (5.5) is in accordance with R in (5.9). In order to formally prove this assertion, we introduce the following assumptions.

Assumption A10. $\widetilde{R}$ *is weakly separable in* $[\mathcal{W}(t), \mathcal{A}(t)]$, *i.e.*

$$\widetilde{R}\big[\ell_r(t), \mathcal{W}(t), \mathcal{A}(t)\big] = \widehat{R}\big[\ell_r(t), Q[\mathcal{W}(t), \mathcal{A}(t)]\big] \;,$$

with $\widehat{R}_Q > 0$.

Assumption A11. Q *is additively separable in all vintages* $v \leq t$, *i.e.*

$$Q\big[\mathcal{W}(t), \mathcal{A}(t)\big] = \int_{-\infty}^{t} F\big[w(v;t), \alpha(v)\big] dv \ ,$$

with $F_w, F_\alpha > 0$.

With the help of these assumptions, we obtain the following representation result.

Proposition 5.1. *Suppose the recycling technology satisfies assumptions A10 and A11. Then*

$$\widetilde{R}\big[\ell_r(t), \mathcal{W}(t), \mathcal{A}(t)\big] = R\big[\ell_r(t), w(t), \alpha_\mu(t), \alpha_\sigma(t)\big] \ ,$$

with $R_w, R_{\alpha_\mu} > 0$ *and* $R_{\alpha_\sigma} \gtreqless 0$, *in general.*

Proof. Applying assumptions A10 and A11 to the general recycling function (5.5) yields

$$\widetilde{R}\big[\ell_r(t), \mathcal{W}(t), \mathcal{A}(t)\big] = \widehat{R}\left[\ell_r(t), \int_{-\infty}^{t} F\big[w(t)f(v;t), \alpha(v)\big] dv\right] \ , \quad (5.10)$$

where $f(v;t) := w(v;t)/w(t)$ with $\int_{-\infty}^{t} f(v;t) dv = 1$ represents the density function of residuals at time t. This density function determines not only the relative frequency of vintage v in time t residuals, but also the relative frequency of recyclability $\alpha(v)$ in time t residuals. In order to obtain the recycling function (5.9), it remains to show that the integral in (5.10) is a function of $w(t)$, $\alpha_\mu(t)$ and $\alpha_\sigma(t)$ only. For that purpose, we introduce the normalization

$$\tilde{\alpha}(v;t) := \frac{\alpha(v) - \alpha_\mu(t)}{\sqrt{\alpha_\sigma(t)}} \ .$$

$\tilde{\alpha}(v;t)$ is the *normalized* recyclability of vintage v residuals at time t. To see this, note that the relative frequency of $\tilde{\alpha}(v;t)$ in time t residuals is the same as the relative frequency of $\alpha(v)$ in time t residuals and, thus, the density function of $\tilde{\alpha}(v;t)$ at time t is $f(v;t)$. The mean of the residuals' normalized recyclability then becomes

$$\tilde{\alpha}_\mu(t) = \int_{-\infty}^{t} \tilde{\alpha}(v;t) f(v;t) dv$$

$$= \frac{1}{w(t)\sqrt{\alpha_\sigma(t)}} \int_{-\infty}^{t} \big[\alpha(v) - \alpha_\mu(t)\big] w(v;t) dv = 0 \ ,$$

and the variance of the residuals' normalized recyclability is

$$\tilde{\alpha}_\sigma(t) = \int_{-\infty}^{t} \left[\tilde{\alpha}(v;t) - \tilde{\alpha}_\mu(t) \right]^2 f(v;t)dv$$

$$= \frac{1}{w(t)\alpha_\sigma(t)} \int_{-\infty}^{t} \left[\alpha(v) - \alpha_\mu(t) \right]^2 w(v;t)dv = 1 \, .$$

Now, using the definition of $\tilde{\alpha}(v;t)$ in (5.10) and taking into account that, at time t, $\tilde{\alpha}(v;t)$ and $f(v;t)$ are given functions of v which vanish through integration establishes

$$\widehat{R}\left[\ell_r(t), \int_{-\infty}^{t} F\left[w(t)f(v;t), \alpha_\mu(t) + \sqrt{\alpha_\sigma(t)}\tilde{\alpha}(v;t) \right] dv \right]$$

$$=: R[\ell_r(t), w(t), \alpha_\mu(t), \alpha_\sigma(t)] \, . \quad (5.11)$$

The partial derivatives of R with respect to $w(t)$ and $\alpha_\mu(t)$ are

$$R_w = \widehat{R}_Q[\cdot] \int_{-\infty}^{t} F_w[\cdot]f(v;t)dv \, , \qquad R_{\alpha_\mu} = \widehat{R}_Q[\cdot] \int_{-\infty}^{t} F_\alpha[\cdot]dv \, .$$

Both derivatives are positive due to $\widehat{R}_Q, F_w, F_\alpha > 0$ and $f(v;t) \geq 0$ with strict inequality for at least one v. It remains to examine the partial derivative of R with respect to $\alpha_\sigma(t)$. From (5.11), we obtain the expression

$$R_{\alpha_\sigma} = \frac{1}{2\sqrt{\alpha_\sigma(t)}} \, \widehat{R}_Q[\cdot] \int_{-\infty}^{t} F_\alpha[\cdot]\tilde{\alpha}(v;t)dv \, , \quad (5.12)$$

which is ambiguous in sign since the normalized recyclability $\tilde{\alpha}(v;t)$ can take positive or negative values. □

Proposition 5.1 identifies conditions (assumptions A10 and A11) under which the general recycling function $\widetilde{R}$ in (5.5) is represented by the function R defined in (5.9). To judge whether the assumptions A10 and A11 are plausible or realistic, first note that the weak separability required in assumption A10 is satisfied by all functional forms typically used in economic models, for example, CES or Cobb-Douglas functions. Moreover, we think that the additive separability invoked in assumption A11 is realistic especially for recycling processes of durable goods like, for example, cars or refrigerators. Such durables usually possess large volume or weight and every worn-out unit needs to be processed separately. There is no mixing of heterogeneous residuals and no mixing of vintages. So, we conclude that assumptions A10 and A11 are plausible for real world recycling processes, and in the subsequent analysis, we therefore restrict our attention to recycling functions of type (5.9).

Efficient Allocation

This paragraph derives the Pareto efficient allocation of the durable good economy described in the previous paragraph. The efficient allocation is obtained by maximizing the present value of the representative consumer's utility

$$\int_0^\infty U\left[c(t), \ell(t)\right] e^{-\delta t} dt$$

subject to (5.1), (5.2), (5.4), (5.6)–(5.9) and the labor supply constraint $\ell(t) = \ell_y(t)+\ell_r(t)$ where $\delta > 0$ is the social discount rate. This maximization problem is an optimal control problem with the integral state constraints in (5.2), (5.6), (5.7) and (5.8). $c(t)$, $w(t)$, $\mu(t)$ and $\sigma(t)$ are the state variables and all other economic variables represent control variables. In order to solve the control problem, we employ the current-value Lagrangean

$$\mathcal{L}(t) = U\left[c(t), \ell(t)\right] + \int_t^\infty D(v - t)y(t)\theta_c(v)e^{-\delta(v-t)}dv$$

$$- \int_t^\infty D_a(v - t)y(t)\theta_w(v)e^{-\delta(v-t)}dv$$

$$- \int_t^\infty D_a(v - t)y(t)\alpha(t)\theta_\mu(v)e^{-\delta(v-t)}dv$$

$$- \int_t^\infty D_a(v - t)y(t)\alpha^2(t)\theta_\sigma(v)e^{-\delta(v-t)}dv$$

$$+ \theta_y(t)\left[Y\left[\ell_y(t), r(t), \alpha(t)\right] - y(t)\right]$$

$$+ \theta_r(t)\left[R\left[\ell_r(t), w(t), \alpha_\mu(t), \alpha_\sigma(t)\right] - r(t)\right] + \theta_{\mu w}(t)\left[\frac{\mu(t)}{w(t)} - \alpha_\mu(t)\right]$$

$$+ \theta_{\sigma w}(t)\left[\frac{\sigma(t)}{w(t)} - \alpha_\mu^2(t) - \alpha_\sigma(t)\right] + \theta_\ell(t)\left[\ell(t) - \ell_y(t) - \ell_r(t)\right],$$

for all $t \in [0, \infty[$. $\theta_c(t)$, $\theta_w(t)$, $\theta_\mu(t)$ and $\theta_\sigma(t)$ are costate variables associated with the state variables $c(t)$, $w(t)$, $\mu(t)$ and $\sigma(t)$, and the other θs are Lagrange multipliers. All θs are interpreted as shadow prices of the respective state or control variable. Therefore, the Lagrangean $\mathcal{L}(t)$ measures the impact of time t control and state variables on efficiency at all times $v \geq t$. We assume that an interior solution exists and restrict our attention to such a solution. Applying the general optimal control results of Bakke (1974) and Kamien and Muller (1976), FOCs for an interior solution are obtained by setting the partial derivatives of $\mathcal{L}(t)$ with respect to all control variables equal to zero, i.e. $\partial\mathcal{L}(t)/\partial\ell(t) = \partial\mathcal{L}(t)/\partial r(t) = \partial\mathcal{L}(t)/\partial\alpha(t) = \ldots = 0$ for all $t \in [0, \infty[$, and the partial derivatives with respect to the state variables equal to the corresponding costate variables, i.e. $\partial\mathcal{L}(t)/\partial c(t) = \theta_c(t)$, $\partial\mathcal{L}(t)/\partial w(t) = \theta_w(t)$,

$\partial \mathcal{L}(t)/\partial \mu(t) = \theta_\mu(t)$ and $\partial \mathcal{L}(t)/\partial \sigma(t) = \theta_\sigma(t)$ for all $t \in [0, \infty[$.[4] To avoid clutter in notation, we define $\psi(v, t) := e^{-\delta(v-t)} U_\ell(v)/U_\ell(t)$ which is a discount factor expressing time v values in terms of time t labor. Rearranging the FOCs for the efficient allocation then yields

Proposition 5.2. *The Pareto efficient allocation is characterized by*

$$
-\frac{Y_\alpha(t)}{y(t)Y_\ell(t)} = -\int_t^\infty D_a(v) \frac{R_{\alpha_\mu}(v)\psi(v,t)}{w(v)R_\ell(v)} dv
$$

$$
- 2\int_t^\infty D_a(v) \frac{R_{\alpha_\sigma}(v)\psi(v,t)}{w(v)R_\ell(v)} \Big[\alpha(t) - \alpha_\mu(v)\Big] dv , \qquad (5.13)
$$

$$
\frac{1}{Y_\ell(t)} = -\int_t^\infty D(v) \frac{U_c(v)\psi(v,t)}{U_\ell(v)} dv - \int_t^\infty D_a(v) \frac{R_w(v)\psi(v,t)}{R_\ell(v)} dv
$$

$$
- \int_t^\infty D_a(v) \frac{R_{\alpha_\sigma}(v)\psi(v,t)}{w(v)R_\ell(v)} \Big[[\alpha(t) - \alpha_\mu(v)]^2 - \alpha_\sigma(v)\Big] dv
$$

$$
- \int_t^\infty D_a(v) \frac{R_{\alpha_\mu}(v)\psi(v,t)}{w(v)R_\ell(v)} \Big[\alpha(t) - \alpha_\mu(v)\Big] dv , \qquad (5.14)
$$

$$
\frac{Y_r(t)}{Y_\ell(t)} = \frac{1}{R_\ell(t)} , \qquad\qquad\qquad (5.15)
$$

for all $t \in [0, \infty[$.

Equation (5.13) represents the allocation rule for efficient recyclability $\alpha(t)$ of the durable good. It says that recyclability should be increased until its marginal production costs (LHS) offset its marginal recycling benefits (RHS). Marginal recycling benefits consist of two components. First, enhancing the durable's recyclability at time t increases the mean recyclability of residuals and the productivity of the recycling process at times $v \geq t$.[5] This effect is formally represented by the term in (5.13) containing the partial derivative R_{α_μ}.

---

[4] Observe that Bakke (1974) and Kamien and Muller (1976) exclusively focus on integral constraints and do not account for 'ordinary' constraints like e.g. (5.1) or (5.9). However, if we assume that the pertinent constraint qualifications are satisfied, then the results of Bakke (1974) and Kamien and Muller (1976) can be extended in the same way as e.g. Seierstad and Sydæster (1987, pp. 269) extend the 'standard' optimal control problem to the case of 'ordinary' constraints. Moreover, Bakke (1974) and Kamien and Muller (1976) employ the present-value Lagrangean instead of its current-value counterpart. But it can be shown that both Lagrangeans imply the same FOCs.

[5] Remember the use of the phrase 'at times $v \geq t$' as explained in footnote 5 on p. 28.

Second, as long as there are still changes in the durable's recyclability over time (so that $\alpha(t)$ is not equal to $\alpha_\mu(v)$ for all $v \geq t$), increasing the durable's recyclability $\alpha(t)$ influences the variance of residuals' recyclability which, in turn, changes the recycling output at times $v \geq t$. This effect is formally captured by the term in (5.13) containing the partial derivative R_{α_σ}. The sign of this term is indeterminate, in general. When it is positive (negative), it represents marginal benefits (costs) of recyclability.

Equation (5.14) is the allocation rule for efficient output $y(t)$ of the durable good. According to this rule, output $y(t)$ has to be increased until its marginal costs offset its marginal benefits. $1/Y_\ell(t)$ equals marginal production costs of $y(t)$, the term containing U_c measures the consumer's marginal willingness-to-pay for $y(t)$ and the term containing R_w represents marginal recycling benefits of $y(t)$ due to an increase in the amount of residuals at times $v \geq t$. Furthermore, a marginal increase in $y(t)$ has two additional effects on recycling through changes in the mean and the variance of the residuals' recyclability at times $v \geq t$. For example, suppose that the durable's recyclability $\alpha(t)$ is greater than the mean recyclability of residuals. Then, increasing the durable's output $y(t)$ gives $\alpha(t)$ more weight in the mean of the residuals' recyclability and raises $\alpha_\mu(v)$ and the amount of recycled material at $v \geq t$. An analogous interpretation holds for the variance of the residuals' recyclability. These recycling effects of $y(t)$ are captured by the terms in (5.14) containing R_{α_μ} and R_{α_σ}. The signs of both terms are indeterminate, in general.

The characterization of the Pareto efficient allocation in our recycling durable good model is completed by equation (5.15). This equation represents the allocation rule for the efficient amount of recycled material $r(t)$ at time t. It requires that marginal labor costs of recycling (RHS) coincide with the marginal productivity of recycled material in the process of producing the durable good (LHS).

We distinguish again between the (short run) adjustment path and the (long run) stationary state of the durable good economy. (5.13)–(5.15) characterize the Pareto efficient allocation on the adjustment path. These efficiency conditions simplify considerably when focusing on the stationary state. As in Chaps. 2–4, we define an (asymptotic) stationary state as a situation in which all economic variables converge to constant levels. By the same arguments as presented on p. 90 for the steady state of the durable good model in Sect. 4.1, if output and recyclability converge to constant levels, then the integral state constraints in (5.2), (5.6), (5.7) and (5.8) turn into

$$c = y\Gamma\,, \qquad w = y\,, \qquad \alpha_\mu = \alpha \qquad \text{and} \qquad \alpha_\sigma = 0\,, \qquad (5.16)$$

where $\Gamma := \int_0^\infty D(a)da > 0$ is the undiscounted use potential of the durable good defined in Sect. 2.1 on p. 16. According to (5.16), the long run stock of the durable good is in fixed proportion to the long run output of the durable good and the long run amount of residuals equals this output. Furthermore, in the long run, every vintage of the durable has the same recyclability implying that the mean of the residuals' recyclability equals this uniform recyclability of the durable good and that the variance of the residuals' recyclability becomes zero. Inserting (5.16) into the efficiency conditions (5.13)–(5.15) yields the long run efficiency conditions

$$-\frac{Y_\alpha}{yY_\ell} = \frac{R_{\alpha_\mu}}{wR_\ell}\Omega \, , \tag{5.17}$$

$$\frac{1}{Y_\ell} = -\frac{U_c}{U_\ell}\Psi + \frac{R_w}{R_\ell}\Omega \, , \tag{5.18}$$

$$\frac{Y_r}{Y_\ell} = \frac{1}{R_\ell} \, , \tag{5.19}$$

where $\Psi := \int_0^\infty D(a)e^{-\delta a}da > 0$ is the discounted use potential of the durable good and where $\Omega := -\int_0^\infty D_a(a)e^{-\delta a}da > 0$. In the long run, the allocation rule for the efficient amount of recycled material remains unchanged as can be seen by comparing (5.15) and (5.19). However, comparing (5.13) and (5.17) elucidates that the marginal recycling effect of the durable's recyclability caused by changes in the variance of the residuals' recyclability vanishes in the steady state since α_σ is always zero and, thus, independent of the durable's recyclability α. Furthermore, the effects of the output y on the mean and variance of the residuals' recyclability vanish in the long run since both variables are independent of output y according to (5.16). Consequently, the accompanying recycling effects of output also vanish as shown by comparing (5.14) and (5.18). These differences between the short run and the long run allocation turn out to be important in the following analysis.

5.1.2 Benchmark Market System

In this section, we introduce a benchmark economy with a set of competitive markets and assume that all markets under consideration operate smoothly. In doing so, we exclusively focus on the case in which the producer sells her products.[6] The aim is to investigate how the efficient allocation can be decentralized by prices. Section 5.1.3 then checks the capacity of tax-subsidy schemes to restore efficiency when markets fail to provide appropriate price signals for recyclability.

In the benchmark economy, the following markets are active. On the labor market, the consumer supplies labor to the durable good producer and to the recycling firm at price $p_\ell(t)$. On the market for recycled material, the recycling firm offers material to the producer of the durable good at price $p_r(t)$. Both $p_\ell(t)$ and $p_r(t)$ are 'conventional' market prices, i.e. the Walrasian auctioneer sets these prices such that demand equals supply of labor and recycled material, respectively. In contrast, on the markets for durables and residuals, prices take the form of hedonic price functions. The durable good

---

[6] In Sect. 5.2, we analyze both sales and rental markets.

producer sells her output $y(t)$ to the consumer at price $P^y[\alpha(t)]$ which depends on the durable's recyclability. The consumer, in turn, offers residuals to the recycler at price $P^w[\alpha_\mu(t), \alpha_\sigma(t)]$ which depends on the mean and variance of the residuals' recyclability. Hence, there are indirect markets for recyclability represented by the derivatives P^y_α, $P^w_{\alpha_\mu}$ and $P^w_{\alpha_\sigma}$. The Walrasian auctioneer aims at determining the price function of the durable good to secure both that the quantity demanded matches the quantity supplied of the durable good and that demand equals supply of the durable's recyclability. The same holds for the price of residuals which has to equate supply and demand for residuals and their recyclability.

In addition to this market transactions, the government imposes particular tax instruments. Although these tax instruments become relevant not until Sect. 5.1.3, we introduce them here in order to avoid repetition in the subsequent analysis. More specifically, the government introduces taxes on output and recyclability of the durable consumption good. The pertinent tax rates at time t are $\tau_y(t)$ and $\tau_\alpha(t)$, respectively. As in the previous chapters, these tax rates are not restricted in sign so that a tax rate represents a tax if it is positive and a subsidy if it is negative. Both tax rates are levied on the durable good producer and we assume that the producer is the only agent who is affected by governmental regulation. Hence, we focus on EPR instruments which, in connection with recycling, sometimes are denoted as upstream regulation (e.g. Calcott and Walls, 1999). As already mentioned in Sect. 4.3, upstream regulation has the advantages of avoiding moonlight dumping and of being independent of the (bounded) rationality of consumers. Furthermore, Calcott and Walls (1999) argue that transaction costs of upstream policies are relatively low since such policies require to regulate a smaller number of agents than downstream regulation.

Table 5.1 summarizes the market transactions (column 1) with the associated market prices (column 2) and the tax rates imposed on the durable good producer (column 3). With this information at hand, we now specify the maximization problems of the market participants. The durable good producer and the recycler maximize the present value of their profits over the planning period $[0, \infty[$ subject to their technologies. The producer faces the production constraint (5.1) whereas the recycler takes into account the recycling function (5.9). The maximization of the present value of profits is equivalent to maximizing instantaneous profits at all times $t \in [0, \infty[$ since neither the producer nor the recycler are subject to dynamic constraints. Therefore, the pertinent Lagrangean of the durable good producer and the recycler read, respectively,

$$\mathcal{L}^y(t) = \left[P^y[\alpha(t)] - \tau_y(t) \right] y(t) - \tau_\alpha(t)\alpha(t) - p_\ell(t)\ell_y(t) - p_r(t)r(t)$$

$$+ \lambda_y(t) \left[Y[\ell_y(t), r(t), \alpha(t)] - y(t) \right], \quad (5.20)$$

Table 5.1. Market Prices and Tax Rates in the Market Economy

marketed goods	market prices	tax rates
1	2	3
labor (ℓ, ℓ_y, ℓ_r)	p_ℓ	–
material (r)	p_r	–
durable good (y)	$P^y(\alpha)$	τ_y
residuals (w)	$P^w(\alpha_\mu, \alpha_\sigma)$	–
recyclability of the durable (α)	–	τ_α

$$\mathcal{L}^r(t) = p_r(t)r(t) - p_\ell(t)\ell_r(t) - P^w\left[\alpha_\mu(t), \alpha_\sigma(t)\right] w(t)$$

$$+ \lambda_r(t)\left[R\left[\ell_r(t), w(t), \alpha_\mu(t), \alpha_\sigma(t)\right] - r(t)\right], \quad (5.21)$$

where $\lambda_y(t)$ and $\lambda_r(t)$ are Langrangean multipliers. The first row in (5.20) equals the producer's instantaneous profit at time t whereas the second row contains the production constraint. Similar, the first row in (5.21) is the recycler's instantaneous profit at time t and the second row reflects the recycling technology. The FOCs for solving the producer's and the recycler's profit maximization problems are obtained by setting equal to zero the partial derivatives of (5.20) and (5.21) with respect to the decision variables and the Lagrangean multipliers.

The consumer maximizes the present value of utility (5.3) subject to her budget constraint and subject to (5.2), (5.4) and (5.6)–(5.8). Like the welfare maximization problem considered in Sect. 5.1.1, this is an optimal control problem with integral state constraints. The current-value Lagrangean is

$$\mathcal{L}^c(t) = U\left[c(t), \ell(t)\right] + \int_t^\infty D(v - t)y(t)\lambda_c(v)e^{-\delta(v-t)}dv$$

$$- \int_t^\infty D_a(v - t)y(t)\lambda_w(v)e^{-\delta(v-t)}dv$$

$$- \int_t^\infty D_a(v - t)y(t)\alpha(t)\lambda_\mu(v)e^{-\delta(v-t)}dv$$

$$- \int_t^\infty D_a(v - t)y(t)\alpha^2(t)\lambda_\sigma(v)e^{-\delta(v-t)}dv + \lambda_{\mu w}(t)\left[\frac{\mu(t)}{w(t)} - \alpha_\mu(t)\right]$$

$$+ \lambda_{\sigma w}(t) \left[\frac{\sigma(t)}{w(t)} - \alpha_\mu^2(t) - \alpha_\sigma(t) \right]$$

$$+ \lambda_\ell(t) \left[p_\ell(t)\ell(t) + P^w[\alpha_\mu(t), \alpha_\sigma(t)]w(t) + x(t) - P^y[\alpha(t)]y(t) \right], \quad (5.22)$$

for all $t \in [0, \infty[$. $x(t)$ is a lump-sum transfer of profits and net tax revenues. $\lambda_c(t)$, $\lambda_w(t)$, $\lambda_\mu(t)$ and $\lambda_\sigma(t)$ are costate variables associated with the state variables $c(t)$, $w(t)$, $\mu(t)$ and $\sigma(t)$ and the other λs are Lagrange multipliers. The last row in (5.22) represents the consumer's budget constraint. It states that the consumer's expenditures for the durable good do not exceed the consumer's income consisting of labor income, income from purchasing consumption residuals and the lump-sum transfer of profits and net tax revenues.

The FOCs for solving the consumer's utility maximization problem are obtained by setting equal to zero the partial derivatives of $\mathcal{L}^c(t)$ with respect to the control variables and by setting equal to the corresponding costate variables the partial derivatives with respect to the state variables. Rearranging and combining the resulting terms with the FOCs of the profit maximization problems of both the producer and the recycler yields

$$-\frac{Y_\alpha(t)}{Y_\ell(t)} = \frac{P_\alpha^y(t)y(t) - \tau_\alpha(t)}{p_\ell(t)}, \quad (5.23)$$

$$\frac{1}{Y_\ell(t)} = \frac{P^y(t) - \tau_y(t)}{p_\ell(t)}, \quad (5.24)$$

$$\frac{Y_r(t)}{Y_\ell(t)} = \frac{p_r(t)}{p_\ell(t)}, \quad (5.25)$$

with

$$\frac{P_\alpha^y(t)}{p_\ell(t)} = -\int_t^\infty D_a(v) \frac{P_{\alpha_\mu}^w(v)\psi(v,t)}{p_\ell(v)} dv$$

$$- 2\int_t^\infty D_a(v) \frac{P_{\alpha_\sigma}^w(v)\psi(v,t)}{p_\ell(v)} \left[\alpha(t) - \alpha_\mu(v) \right] dv, \quad (5.26)$$

$$\frac{P^y(t)}{p_\ell(t)} = -\int_t^\infty D(v) \frac{U_c(v)\psi(v,t)}{U_\ell(v)} dv - \int_t^\infty D_a(v) \frac{P^w(v)\psi(v,t)}{p_\ell(v)} dv$$

$$- \int_t^\infty D_a(v) \frac{P_{\alpha_\sigma}^w(v)\psi(v,t)}{p_\ell(v)} \left[\left[\alpha(t) - \alpha_\mu(v) \right]^2 - \alpha_\sigma(v) \right] dv$$

$$- \int_t^\infty D_a(v) \frac{P_{\alpha_\mu}^w(v)\psi(v,t)}{p_\ell(v)} \Big[\alpha(t) - \alpha_\mu(v)\Big] dv \,, \tag{5.27}$$

$$\frac{P^w(t)}{p_\ell(t)} = \frac{R_w(t)}{R_\ell(t)} \,, \quad \frac{P_{\alpha_\mu}^w(t)}{p_\ell(t)} = \frac{R_{\alpha_\mu}(t)}{w(t)R_\ell(t)} \,, \tag{5.28}$$

$$\frac{P_{\alpha_\sigma}^w(t)}{p_\ell(t)} = \frac{R_{\alpha_\sigma}(t)}{w(t)R_\ell(t)} \,, \quad \frac{p_r(t)}{p_\ell(t)} = \frac{1}{R_\ell(t)} \,, \tag{5.29}$$

for all $t \in [0, \infty[$. Note that all prices in (5.23)–(5.29) are divided by the price of labor. Without loss of generality, we therefore choose labor as numeraire and set $p_\ell(t) = 1$ for all $t \in [0, \infty[$. The FOCs (5.23)–(5.29) together with the constraints of the profit and utility maximization problems (5.20)–(5.22) determine a time path of the market equilibrium for the interval $t \in [0, \infty[$. The equilibrium is assumed to exist and to be unique.

To see how our benchmark economy performs, suppose the government abstains from imposing any taxes and subsidies, insert (5.26)–(5.29) into the market equilibrium conditions (5.23)–(5.25) and compare the resulting expressions with the efficiency conditions (5.13)–(5.15). We then obtain

Proposition 5.3. *Suppose all markets of the benchmark economy work and set* $\tau_y(t) = \tau_\alpha(t) = 0$ *for all* $t \in [0, \infty[$. *Then the market equilibrium is Pareto efficient at all times* $t \in [0, \infty[$.

According to Proposition 5.3, the benchmark economy is capable of implementing the Pareto efficient allocation. Equations (5.26)–(5.29) provide information about how prices guide the allocation towards efficiency. The price of the durable good, $P^y(t)$, equals the consumer's marginal willingness-to-pay for the output of the durable plus a price mark-up. This mark-up reflects the consumer's revenues from selling residuals of vintage t and the impact of vintage t output on these revenues caused by changes in the mean and variance of the residuals' recyclability.[7] The indirect price for the durable's recyclability, $P_\alpha^y(t)$, the price of residuals, $P^w(t)$, the indirect price of the

---

[7] In the partial equilibrium models of Chaps. 2–4, we proceeded on the assumptions (a) that the consumers' willingness-to-pay for the durable good equals the present value of the consumer's willingness-to-pay for the services the durable generates during its whole lifetime (for example, on p. 32) and (b) that rational utility-maximizing consumers adjust their willingness-to-pay for the durable good by waste taxes paid at the end of the product's life (for example, on p. 72). These assertions become obvious from (5.27) by observing that $P^y(t)$ is the consumer's willingness-to-pay for the durable good at time t, that $-U_c(v)/U_\ell(v)$ is the consumer's willingness-to-pay for the durable's services at time v and that $P^w(v)$ can be interpreted as tax imposed on consumption residuals or, equivalently, on consumption waste at time v. The same argument holds for the statement that consumers adjust their willingness-to-pay for the durable good by stock taxes paid during the life of the consumption good.

mean of the residuals' recyclability, $P_{\alpha_\mu}^w(t)$, and the indirect price of the variance of the residuals' recyclability, $P_{\alpha_\sigma}^w(t)$, equal marginal recycling benefits of the respective variable. Finally, the price of recycled material, $p_r(t)$, reflects marginal labor costs of recycling. In sum, all marginal costs and benefits of output, recyclability and recycled material are reflected in market prices in an appropriate way, and this is why markets are capable to support the efficient allocation.

5.1.3 Market Failure and its Correction

At this point, it is natural to ask whether the markets of our benchmark economy can be expected to work in real economies. Quite obviously, the labor market, the market for recycled material and the market for the output of the durable good do exist in the real world. Thus, potential candidates for missing markets are the market for residuals and the indirect markets for recyclability. Following Eichner and Pethig (2001) and Calcott and Walls (2000), we restrict our investigations to missing price signals for recyclability. This case is of particular interest since product attributes are public goods[8] and markets for public goods usually do not emerge in reality.

When markets for recyclability fail to be active, then $P_\alpha^y(t) = P_{\alpha_\mu}^w(t) = P_{\alpha_\sigma}^w(t) = 0$ for all $t \in [0, \infty[$ and the marginal effects of recyclability and output reflected by these indirect prices are no longer contained in the market price system. Consequently, the missing markets create a recyclability and an output externality since the producer of the durable good sets recyclability and output without taking into account the impact on the recycling process. This market failure can be corrected by setting appropriate rates of the policy instruments $\tau_y(t)$ and $\tau_\alpha(t)$. Comparing the efficiency conditions (5.13)–(5.15) with the market equilibrium conditions (5.23)–(5.29) immediately yields

Proposition 5.4. *Suppose there are no markets for recyclability, i.e.* $P_\alpha^y(t) = P_{\alpha_\mu}^w(t) = P_{\alpha_\sigma}^w(t) = 0$ *for all* $t \in [0, \infty[$ *and set*

$$\tau_\alpha(t) = \int_t^\infty D_a(v) \frac{R_{\alpha_\mu}(v)\psi(v,t)}{w(v)R_\ell(v)} dv$$

$$+ 2 \int_t^\infty D_a(v) \frac{R_{\alpha_\sigma}(v)\psi(v,t)}{w(v)R_\ell(v)} \Big[\alpha(t) - \alpha_\mu(v)\Big] dv \, ,$$

$$\tau_y(t) = - \int_t^\infty D_a(v) \frac{R_{\alpha_\mu}(v)\psi(v,t)}{w(v)R_\ell(v)} \Big[\alpha(t) - \alpha_\mu(v)\Big] dv$$

---

[8] In order to reveal the public good property of recyclability, imagine a disaggregated version of our model with many recycling firms. Since the products of a durable good producer are usually recycled by more than one recycling firm, every recycler 'consumes' the same recyclability. According to Blümel et al. (1986), joint consumption is the constitutive property of public goods.

$$- \int_t^\infty D_a(v) \frac{R_{\alpha_\sigma}(v)\psi(v,t)}{w(v)R_\ell(v)} \left[[\alpha(t) - \alpha_\mu(v)]^2 - \alpha_\sigma(v) \right] dv \; ,$$

for all $t \in [0,\infty[$. Then the market allocation is Pareto efficient.

Without markets for recyclability and without any policy intervention, the durable good producer does not obtain price signals related to recyclability of the durable. It is then sensible for the producer to ignore the impact the durable's recyclability exerts on the mean and variance of the residuals' recyclability and on the recycling process. This constitutes the recyclability externality since the durable good producer determines $\alpha(t)$ irrespective of the recycler's needs or wants. According to Proposition 5.4, the recyclability externality is internalized by a recyclability tax or subsidy $\tau_\alpha(t)$ which is the usual Pigouvian policy instrument: It reflects the marginal recycling benefits/costs of the durable's recyclability caused by changes in the mean and variance of residuals' recyclability. It induces the durable good producer to account for the impact of her product design on recycling and thus results in an efficient level of recyclability. In general, $\tau_\alpha(t)$ is ambiguous in sign such that it may represent a subsidy ($\tau_\alpha(t) < 0$) or a tax ($\tau_\alpha(t) > 0$).

On its own, $\tau_\alpha(t)$ is not able to bridge the inefficiency gap in case that markets for recyclability fail. According to Proposition 5.4, the second policy instrument $\tau_y(t)$ is indispensable. The intuition is as follows. Without price signals $P_\alpha^y(t)$, $P_{\alpha_\mu}^w(t)$ and $P_{\alpha_\sigma}^w(t)$ there exists not only a recyclability externality, but also an output externality since the producer's choice of the output $y(t)$ affects the mean and the variance of the residuals' recyclability, too. As can be seen from (5.27), in the benchmark economy, this impact of output on the residuals' recyclability is reflected by the mark-up containing $P_{\alpha_\mu}^w$ and $P_{\alpha_\sigma}^w$ in the durable's price P^y. But when price signals for recyclability are missing, then the mark-up vanishes and the price of the durable does not reflect full social costs and benefits. The durable good producer does not take into account the entire impact of her output on the recycling process and, hence, she creates an output externality. The regulator then needs the additional instrument $\tau_y(t)$ in order to internalize this output externality. Like the sign of $\tau_\alpha(t)$, also the sign of $\tau_y(t)$ is indeterminate, in general.

In the stationary state, the efficiency restoring tax-subsidy scheme suggested in Proposition 5.4 reduces to a much simpler one. Owing to $\alpha_\mu = \alpha$ and $\alpha_\sigma = 0$ from (5.16) we get

$$\tau_\alpha = -\frac{R_{\alpha_\mu}}{wR_\ell}\Omega < 0 \qquad \text{and} \qquad \tau_y = 0 \; . \tag{5.30}$$

In the long run, efficiency requires only a recyclability subsidy. There are two reasons for this result. First, the durable's recyclability does not influence the variance of the residuals' recyclability anymore since this variance is zero. The associated marginal recycling effect vanishes and τ_α becomes a subsidy because it reflects the marginal recycling benefits of the mean of the residual's recyclability only. Second, the output of the durable good does not exert an effect on the mean and the variance of the residuals' recyclability in the long run since α_μ is equal to α and α_σ is zero. Hence, the output externality completely vanishes and an output tax or subsidy is therefore not needed anymore.

The studies using static general equilibrium models which we already referred to in the introduction to this chapter concluded that a subsidy on recyclability is sufficient to restore efficiency in case that markets for recyclability are absent. Our results in Proposition 5.4 and equation (5.30) reveal that this is true only in economies with nondurable consumption goods (as is implicitly presupposed in those studies) or in the steady state of economies with durable consumption goods. The efficient policy approach on the adjustment path of a durable good economy, however, needs to employ an additional tax or subsidy on the output of the durable good since this output creates an externality in the recycling sector, too. Furthermore, we gain the insight that in the short run, $\tau_\alpha(t)$ is indeterminate in sign, i.e. on the adjustment path the recyclability subsidy may turn into a recyclability tax.

It should be noted that previous studies do not suggest a recyclability subsidy only, but also derive alternative tax-subsidy schemes which are able to overcome market failure in case that markets for recyclability fail. For example, Eichner and Pethig (2001) show that the recyclability subsidy can be replaced by a subsidy on material input and a tax on output of the production process. Of course, these alternative regulatory schemes have also to be modified when restoring efficiency on the adjustment path of a durable good economy is at issue. The pertinent policy instruments then have to account for the additional externalities created by the output and recyclability of the durable good.

5.2 Durable Goods Markets with Endogenous Product Durability

5.2.1 Model and Efficient Allocation

In the previous section, we have supposed product durability is exogenously given. In the present section, this restriction is dropped.[9] We will now first describe the model and then characterize the efficient allocation under different kinds of time preferences.

Model Description

At time t, the model is given by:[10]

---

[9] The analysis in this section is based on Eichner and Runkel (2003).

[10] To avoid confusion in notation, in this section, we explicitly distinguish between quantities supplied (output) and demanded (input), respectively. The former are indicated by the superscripts s the latter by the superscript d.

$$z^s(t) \leq Z[\underset{+}{\ell_z^d(t)}] \; , \qquad \text{virtual material production} \qquad (5.31)$$

$$y^s(t) \leq Y[\underset{+}{\ell_y^d(t)}, \underset{+}{m^d(t)}] \; , \qquad \text{production of the durable good} \qquad (5.32)$$

$$q_y^s(t) = \frac{m^d(t)}{y^s(t)} \; , \qquad \text{recyclability of the durable good} \qquad (5.33)$$

$$\phi^s(t) = \Phi[\underset{+}{q_y^s(t)}] \; , \qquad \text{durability of the durable good} \qquad (5.34)$$

$$c(t) = \int_{-\infty}^{t} D[t - v, \phi^d(v)]y^d(v)dv \; , \qquad \text{stock of the durable good} \qquad (5.35)$$

$$u(t) \leq U[\underset{+}{c(t)}, \underset{-}{\ell^s(t)}] \; , \qquad \text{household's utility} \qquad (5.36)$$

$$w^s(t) = -\int_{-\infty}^{t} D_a[t - v, \phi^d(v)]y^d(v)dv \; , \qquad \text{amount of residuals} \qquad (5.37)$$

$$b(t) = -\int_{-\infty}^{t} D_a[t - v, \phi^d(v)]y^d(v)q_y^d(v)dv \; , \qquad \text{embodied material of residuals} \qquad (5.38)$$

$$q_r^s(t) = \frac{b(t)}{w^s(t)} \; , \qquad \text{mean of residuals' recyclability} \qquad (5.39)$$

$$r^s(t) \leq R[\underset{+}{\ell_r^d(t)}, \underset{+}{w^d(t)}, \underset{+}{q_r^d(t)}] \; , \qquad \text{recycling technology} \qquad (5.40)$$

$$z^s(t) + r^s(t) \geq m^d(t) \; , \quad y^s(t) \geq y^d(t) \; ,$$

$$w^s(t) \geq w^d(t) \; , \quad \ell^s(t) \geq \ell_z^d(t) + \ell_y^d(t) + \ell_r^d(t) \; ,$$

$$\phi^s(t) \geq \phi^d(t) \; , \quad q_y^s(t) \geq q_y^d(t) \; , \quad q_r^s(t) \geq q_r^d(t) \; , \qquad \text{resource constraints} \qquad (5.41)$$

For interpreting (5.31)–(5.41), suppose the economy under consideration uses and ultimately disposes of one type of virgin material (or an aggregate of different materials with fixed material mix). According to (5.31), virgin material is extracted at time t in quantity $z^s(t)$ with the help of labor input $\ell_z^d(t)$. In the production sector described by (5.32), the output $y^s(t)$ of a durable good (measured in quantity units) is produced with labor $\ell_y(t)$ and material $m^d(t)$ (measured in kg or tons) which consists of virgin material $z^s(t)$ and recycled material $r^s(t)$ as shown in the resource constraint $z^s(t) + r^s(t) \geq m^d(t)$ contained in (5.41).

The modeling of product characteristics differs from that employed in the model of Sect. 5.1 in two respects. First, product recyclability is not specified as an ad hoc variable possessing different interpretations. Instead, we suppose the production process allows for varying the weight per unit, $q_y^s(t)$, defined in (5.33) as the amount of material per unit of the durable good. Product recyclability is set equal to product weight since this weight directly enters the recycling technology in (5.40). Second, we now assume that product durability is not exogenously given, but depends on recyclability according to (5.34). As argued already in the introduction, it is plausible that both recyclability and durability are positively correlated with product weight and, hence, an increase in recyclability at time t brings about an increase in product durability at time t (i.e. $\Phi_q > 0$).

The stock $c(t)$ of the durable good at time t is represented by equation (5.35). As in the previous chapters, it equals the sum of the remaining units of all vintages $v \leq t$. The decay function D is assumed to satisfy the assumptions A1 and A2 on p. 15 so that we are allowed to apply the results derived in Lemma 2.1 on p. 16. At time t, the utility of the representative household in (5.36) is influenced by the durable's services the amount of which is proportional to the physical stock $c(t)$. Furthermore, the consumer suffers an utility loss from labor it supplies.

After consumption at time t, the amount $-D_a[t - v, \phi^d(v)]$ of vintage $v \leq t$ wears out and is turned into consumption residuals. The total amount of residuals at time t is therefore $w^s(t)$ defined in (5.37). $b(t)$ from (5.38) is the material embodied in residuals at time t. It equals total weight or, equivalently, 'total recyclability' of consumption residuals. According to (5.39), dividing $b(t)$ by the amount of residuals yields the (weighted) mean of the residuals' recyclability at time t, $q_r^s(t)$. Inputs of the recycling process (5.40) are the mean of the residuals' recyclability, $q_r^d(t)$, the quantity of consumption residuals, $w^d(t)$, and labor, $\ell_r^d(t)$. Recycling generates $r^s(t)$ units of recycled material to be (re-)used in production.[11] The description of the model is completed by listing the resource constraints in (5.41). These constraints state that the demand of an economic variable must not exceed the supply of this variable.

In the economy (5.31)–(5.41), producers sell their output. Hence, in the economy, a particular property rights regime prevails defined by a specific sequence of transactions: At time t, the production sector supplies the bundle $[y^s(t), q_y^s(t), \phi^s(t)]$ to the household who demands $[y^d(t), q_y^d(t), \phi^d(t)]$ and who,

---

[11] In contrast to the model in Sect. 5.1, we do not assume that the variance of the residuals' recyclability is an argument of the recycling function to keep the analysis in the present section tractable. This assumption is in accordance with the representation theorem of Proposition 5.1 if the function F in assumption A11 is linear in recyclability.

in turn, offers the bundle $[w^s(t), q^s_r(t)]$ to the recycling sector whose demand is $[w^d(t), q^d_r(t)]$. Thus, when purchasing the durable good, the household acquires the property rights for both consumption goods and residuals. This assignment of property rights will be referred to as the *sales rule* (S-rule). In Sect. 5.2.3, we also consider the rental case represented by a different property rights regime which we will call the *rental rule* (R-rule). Under this regime, both consumption goods and residuals remain the property of the production firm.

Efficient Allocation

As a point of reference, we derive the efficient allocation of the durable good economy (5.31)–(5.41). For this, consider a social planner who aims at solving the problem

$$\max \int_0^\infty U[c(t), \ell^s(t)]\rho(t)dt \tag{5.42}$$

subject to (5.31)–(5.35), (5.37)–(5.41). In order to analyze intergenerational aspects of durable consumption goods, we employ, alternatively, two different kinds of time preferences specified by different functional forms of the discount factor $\rho(t)$. In the first case, it is assumed that $\rho(t) = e^{-\delta t}$ with $\delta > 0$. In this conventional case, the objective function in (5.42) represents Utilitarian preferences. In the second case, we suppose $\rho(t) = e^{-\delta(t)t}$ with $\delta(t) > 0$ for all $t \in [0, \infty[$ and $\lim_{t \to \infty} \delta(t) = 0$. The discount rate is no longer constant, but converges to zero as time tends to infinity. According to Beltratti et al. (1997), under this functional form of the discount factor, the objective function in (5.42) satisfies a set of axioms introduced by Chichilnisky (1996). Therefore, we refer to this specification as Chichilnisky time preference. For notational convenience, we call the solution to (5.42) under the Utilitarian time preference *u-efficient* and the solution to (5.42) under the Chichilnisky time preference *c-efficient*.[12]

For either functional form of the discount factor, problem (5.42) is an optimal control problem with integral state constraints (5.35), (5.37) and (5.38). $c(t)$, $w^s(t)$ and $b(t)$ are state variables whereas all other economic variables are controls. In order to derive a solution to this problem, we employ the current-value Lagrangean

$$\mathcal{L}(t) = U[c(t), \ell^s(t)] - \int_t^\infty D_a[v - t, \phi^d(t)]y^d(t)q^d_y(t)\theta_b(v)\rho(v,t)dv$$

---

[12] It should be noted that there are empirical and theoretical pros and cons for both kinds of time preferences. For example, declining discount rates are empirically evident, but may lead to time inconsistency. For further discussion, see Chichilnisky et al. (1997). It is beyond the scope of the present analysis to contribute to this discussion. We simply apply both concepts, point out the differences in their implications for our durable good economy with recycling and leave open the question of the comparative merits of either objective function.

$$+ \int_t^\infty D[v - t, \phi^d(t)] y^d(t) \theta_c(v) \rho(v, t) dv$$

$$- \int_t^\infty D_a[v - t, \phi^d(t)] y^d(t) \theta_w(v) \rho(v, t) dv$$

$$+ \theta_y(t) \left[Y[\ell_y^d(t), m^d(t)] - y^s(t) \right]$$

$$+ \theta_r(t) \left[R[\ell_r^d(t), w^d(t), q_r^d(t)] - r^s(t) \right]$$

$$+ \theta_z(t) \left[Z[\ell_z^d(t)] - z^s(t) \right] + \theta_{qy}(t) \left[\frac{m^d(t)}{y^s(t)} - q_y^s(t) \right]$$

$$+ \theta_\phi(t) \left[\Phi[q_y^s(t)] - \phi^s(t) \right] + \theta_{qr}(t) \left[\frac{b(t)}{w^s(t)} - q_r^s(t) \right]$$

$$+ \theta_m(t) \left[z^s(t) + r^s(t) - m^d(t) \right] + \theta_{ys}(t) \left[y^s(t) - y^d(t) \right]$$

$$+ \theta_{\phi s}(t) \left[\phi^s(t) - \phi^d(t) \right] + \theta_{qs}(t) \left[q_y^s(t) - q_y^d(t) \right]$$

$$+ \theta_{ws}(t) \left[w^s(t) - w^d(t) \right] + \theta_{qq}(t) \left[q_r^s(t) - q_r^d(t) \right]$$

$$+ \theta_\ell(t) \left[\ell^s(t) - \ell_z^d(t) - \ell_y^d(t) - \ell_r^d(t) \right], \tag{5.43}$$

with $\rho(v, t) := \rho(v)/\rho(t)$. θ_c, θ_w and θ_b represent costate variables associated with c, w^s and b, respectively. All other θs are Lagrange multipliers. Following Bakke (1974) and Kamien and Muller (1976), the FOCs for an interior solution are obtained by setting all partial derivatives of (5.43) with respect to the control variables equal to zero, i.e. $\partial \mathcal{L}(t)/\partial \ell_z^d(t) = \partial \mathcal{L}(t)/\partial z^s(t) = \ldots = 0$ for all $t \in [0, \infty[$, and the partial derivatives with respect to the state variables equal to the corresponding costate variables, i.e. $\partial \mathcal{L}(t)/\partial c(t) = \theta_c(t)$, $\partial \mathcal{L}(t)/\partial w^s(t) = \theta_w(t)$ and $\partial \mathcal{L}(t)/\partial b(t) = \theta_b(t)$ for all $t \in [0, \infty[$. As in Sect. 5.1, $\psi(v, t) := U_\ell(v) \rho(v, t)/U_\ell(t)$ is a discount factor expressing time v values in terms of time t labor. The FOCs for the efficient allocation may then be rearranged to yield

Proposition 5.5. *The efficient allocation is characterized by*

$$\underbrace{\frac{1}{Z_\ell(t)} - \frac{Y_m(t)}{Y_\ell(t)}}_{\text{MPC}_q} + \underbrace{\Phi_q(t) \int_t^\infty D_{a\phi}(v) \frac{R_w(v)\psi(v,t)}{R_\ell(v)} dv}_{\text{MRC}_\phi^w} =$$

$$\underbrace{-\Phi_q(t) \int_t^\infty D_\phi(v) \frac{U_c(v)\psi(v,t)}{U_\ell(v)} dv}_{\text{MHU}_\phi} - \underbrace{\int_t^\infty D_a(v) \frac{R_q(v)\psi(v,t)}{w(v)R_\ell(v)} dv}_{\text{MRB}_q^q}$$

$$\underbrace{-\Phi_q(t) \int_t^\infty D_{a\phi}(v) \Big[q_y(t) - q_r(v) \Big] \frac{R_q(v)\psi(v,t)}{w(v)R_\ell(v)} dv}_{\text{MRB}_\phi^q}, \qquad (5.44)$$

and

$$\underbrace{\frac{1}{Y_\ell(t)} + q_y(t) \left[\frac{1}{Z_\ell(t)} - \frac{Y_m(t)}{Y_\ell(t)} \right]}_{\text{MPC}_y} =$$

$$\underbrace{- \int_t^\infty D(v) \frac{U_c(v)\psi(v,t)}{U_\ell(v)} dv}_{\text{MHU}_y} - \underbrace{\int_t^\infty D_a(v) \frac{R_w(v)\psi(v,t)}{R_\ell(v)} dv}_{\text{MRB}_y^w}$$

$$\underbrace{+ \int_t^\infty D_a(v) \Big[q_r(v) - q_y(t) \Big] \frac{R_q(v)\psi(v,t)}{w(v)R_\ell(v)} dv}_{\text{MRB}_y^q}, \qquad (5.45)$$

for all $t \in [0, \infty[$.

Equation (5.44) in Proposition 5.5 is the allocation rule for the efficient recyclability $q_y(t)$ of the durable good. It states that at every time t, marginal social costs of the durable's recyclability equal marginal social benefits. Marginal costs and benefits consist of five components which are illustrated in Fig. 5.1a: First, the first two terms on the LHS of (5.44) represent the (net) marginal production costs of the durable's recyclability (MPC$_q$): For given output $y(t)$, increasing the recyclability requires substituting material for labor. This substitution causes extra material extraction costs $1/Z_\ell(t)$, on the one hand, and saves labor costs $Y_m(t)/Y_\ell(t) = [d\ell_y(t)/dm(t)]|_{y(t)=const.}$ in the production sector, on the other hand. Second, the integral on the LHS of (5.44) represents marginal recycling costs/benefits of the good's durability due to changes in the amount of residuals (MRC$_\phi^w$). It is part of the marginal effects of recyclability since a marginal increase in the durable's recyclability enhances

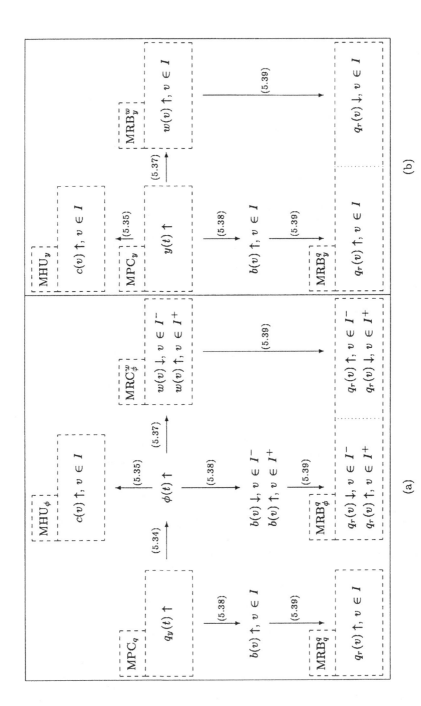

Figure 5.1. Partial Effects of the Durable's (a) Recyclability and (b) Output

product durability. The sign of the integral is indeterminate, in general, since it is based on the waste-delaying effect of durability,[13] i.e. an increase in the good's durability (caused by a shift in the good's recyclability) ceteris paribus reduces the amount of residuals and the recycling output at times $v \in I^- := [t, A(\phi(t))[$, but raises the amount of residuals and the recycling output at times $v \in I^+ :=]A(\phi(t)), \infty[$. Third, the first integral on the RHS of (5.44) is the marginal household utility of durability (MHU$_\phi$): Raising durability in t by increasing the good's recyclability enhances the durable stock $c(v)$ at times $v \in I := [t, \infty[$. Fourth, the second integral on the RHS of (5.44) represents the marginal recycling benefits of the durable's recyclability due to an increase in the residuals' recyclability (MRB$_q^q$): Enhancing the durable's recyclability in t increases embodied material of residuals which, in turn, increases the residuals' recyclability at times $v \in I$. Fifth, the third integral on the RHS of (5.44) represents the marginal recycling benefits/costs of durability due to changes in the residuals' recyclability (MRB$_\phi^q$). In general, this marginal effect is ambiguous in sign since residuals' recyclability is influenced through different channels: On the one hand, increasing durability in t decreases (increases) the embodied material of residuals which, in turn, exerts a negative (positive) effect on the residuals' recyclability at times $v \in I^-$ (at times $v \in I^+$). On the other hand, the improvement of durability decreases (increases) the amount of residuals and hence increases (decreases) the residuals' recyclability at times $v \in I^-$ (at times $v \in I^+$).

Equation (5.45) in Proposition 5.5 represents the allocation rule for efficient output $y(t)$. It states that marginal social costs of output at time t equal marginal social benefits. Marginal costs and marginal benefits comprise four components which are shown in Fig. 5.1b: First, the LHS of (5.45) captures the marginal production costs of output (MPC$_y$) consisting of marginal labor costs $1/Y_\ell(t)$ and (net) marginal material costs $q_y(t)[1/Z_\ell(t) - Y_m(t)/Y_\ell(t)]$. Second, the first integral on the RHS of (5.45) represents marginal household utility of the output (MHU$_y$): Increasing output at time t enhances the durable stock $c(v)$ at times $v \in I$. Third, the second integral on the RHS of (5.45) stands for the marginal recycling benefits of the output due to an increasing amount of residuals (MRB$_y^w$): Raising output at t exerts a positive effect on residuals $w(v)$ at times $v \in I$. Fourth, the third integral on the RHS of (5.45) represents the marginal recycling benefits/costs of the durable good due to changes in the residuals' recyclability (MRB$_y^q$). This integral is indeterminate in sign since the residuals' recyclability is influenced by the output through two channels. On the one hand, if output at time t increases, then embodied material and recyclability of residuals increase at times $v \in I$, too. Second, raising output increases the amount of residuals which, in turn, negatively affects residuals' recyclability at $v \in I$.

---

[13] Strictly speaking, in the model of the present section, durability causes a 'residual-delaying' effect. For simplicity, however, we keep using the term waste-delaying effect.

This interpretation of the optimum conditions (5.44) and (5.45) is valid for the u-efficient and for the c-efficient allocation. The only difference between both allocations is that the present values of the pertinent marginal costs and marginal benefits are computed with different discount rates. The efficiency conditions simplify considerably if we focus on the stationary state of the economy. As in the previous chapters, we define an (asymptotic) stationary state as a situation in which all economic variables converge to constant levels. To characterize such a steady state, suppose output and recyclability of the durable good converge to constant levels. Then, by the same arguments as presented in Sect. 4.1 on p. 90, equations (5.33), (5.35), (5.37), (5.38) and (5.39) imply

$$c = y\Gamma \,, \qquad w = y \,, \qquad q_r = q_y \quad \text{and} \quad b = q_y y = m \,, \qquad (5.46)$$

where $\Gamma(\phi) := \int_0^\infty D(a, \phi)da$ is the undiscounted use potential of the durable good already introduced in Sect. 2.1 on p. 16. According to (5.46), in the long run, the durable stock is proportional to the output of the durable good, the amount of residuals equals this output, the residuals' recyclability equals the durable's recyclability and the material embodied in the residuals equals the material input of the durable good production. To investigate the long run implications for the efficient allocation, consider first the Utilitarian case with $\rho(t) = e^{-\delta t}$. From (5.44) and (5.45) we obtain

Proposition 5.6. (*i*) *The stationary state of the u-efficient allocation is characterized by*

$$\frac{1}{Z_\ell} - \frac{Y_m}{Y_\ell} - \Phi_q \Omega' \frac{R_w}{R_\ell} = -\Phi_q \Psi' \frac{U_c}{U_\ell} + \Omega \frac{R_q}{yR_\ell} \,, \qquad (5.47)$$

$$\frac{1}{Y_\ell} + q_y \left[\frac{1}{Z_\ell} - \frac{Y_m}{Y_\ell} \right] = -\Psi \frac{U_c}{U_\ell} + \Omega \frac{R_w}{R_\ell} \,. \qquad (5.48)$$

(*ii*) *Suppose $R_q > 0$ and consider the assumptions*

Z *is linear:* $Z(\ell_z) = \bar{z}\ell_z$ *with* $\bar{z} > 0 \,,$ (5.49)

Y *is linear homogeneous:* $Y(\varsigma\ell_y, \varsigma m) = \varsigma Y(\ell_y, m)$ *for all* ς, (5.50)

U *is linear:* $U(c, \ell^s) = \bar{c}c - \bar{\ell}\ell^s$ *with* $\bar{c}, \bar{\ell} > 0 \,,$ (5.51)

R *is linear:* $R(\ell_r, w, q_r) = \tilde{\ell}\ell_r + \bar{w}w + \bar{q}q_r$ *with* $\tilde{\ell}, \bar{w}, \bar{q} > 0 \,,$ (5.52)

Ω *and* Ψ *are concave:* $\Omega'', \Psi'' < 0 \,,$ (5.53)

Φ *is linear:* $\Phi(q_y) = \bar{\phi}q_y$ *with* $\bar{\phi} > 0 \,.$ (5.54)

If (5.49)–(5.54) are satisfied, then the u-efficient long run recyclability and the u-efficient long run durability are greater than they would be in case of $R_q \equiv 0$.

Proof. Proposition 5.6i follows directly from (5.44) and (5.45) if we take into account $\rho(t) = e^{-\delta t}$ and the definitions $\Psi(\phi) := \int_0^\infty D(a, \phi)e^{-\delta a}da$ and $\Omega(\phi) := -\int_0^\infty D_a(a, \phi)e^{-\delta a}da$ from Chap. 2. Under the assumptions (5.49)–(5.54), equation (5.47) may be rewritten as

$$F(q_y) := 1/\bar{z} - G(q_y) - H(q_y) = \Omega\bar{q}/y\tilde{\ell}, \qquad (5.55)$$

with $G(q_y) := Y_m/Y_\ell$ and $H(q_y) := \bar{\phi}\Omega'\bar{w}/\tilde{\ell} + \bar{\phi}\Psi'\bar{c}/\ell$. Assumption (5.50) ensures that there exists a function $Q(k) := Y(k, 1) = 1/q_y$ with $k := \ell_y/m$. Consequently, we obtain $Y_\ell = Q_k$, $Y_m = Q - kQ_k$ and $Y_{\ell k} = Q_{kk} = Y_{\ell\ell} < 0$, $Y_{mk} = -kQ_{kk} > 0$. The signs of the second derivatives of Y with respect to k imply that G is increasing in k and decreasing in q_y since $k = Q^{-1}(q_y)$. Furthermore, owing to the concavity of Ω and Ψ we obtain $H_q := \bar{\phi}^2\Omega''\bar{w}/\tilde{\ell} + \bar{\phi}^2\Psi''\bar{c}/\bar{\ell} < 0$. In sum, under (5.49)–(5.54), the function $F(q_y)$ is increasing in q_y. Observing that $R_q \equiv 0$ is equivalent to $\bar{q} = 0$ proves Proposition 5.6ii. $\square$

(5.47) and (5.48) are the allocation rules for u-efficient recyclability and u-efficient output of the durable good in the long run. Their interpretation is analogous to that of the short run allocation rules (5.44) and (5.45) with two simplifications: First, the MRB$^q_\phi$ from (5.44) and the MRB^q_y from (5.45) are zero since the residuals' long run recyclability equals the durable's long run recyclability and does not depend on durability or output according to (5.46). Second, the MRC$^w_\phi$ from (5.44) unambiguously represent marginal costs since they simplify to $-\Phi_q\Omega'R_w/R_\ell > 0$ in the long run. It is important to note that this simplification is solely due to the constant and positive discount rate used in deriving the u-efficient allocation: The MRC$^w_\phi$ are based on the waste-delaying effect of enhancing durability (caused by an increase in the durable's recyclability). This effect reduces the amount of residuals and the recycling output in early periods while raising the amount of residuals and the recycling output at later dates. In the long run, the changes in recycling output cancel out since the total amount of residuals is not influenced by the increase in durability and since marginal recycling productivity of residuals measured by R_w/R_ℓ is constant over time. Nevertheless, due to the positive discount rate society prefers recycling output at early dates and, consequently, the waste-delaying effect causes social costs.

Proposition 5.6ii shows that under some qualifications the u-efficient long run recyclability and durability are greater in case of $R_q > 0$ than they would be if R_q were zero. Of course, the conditions under which this result holds are quite restrictive, but the result has a strong economic intuition since for $R_q > 0$ increasing recyclability has the additional benefits MRB^q of increasing recycling productivity. Therefore, we conjecture that Proposition 5.6ii is also valid under more general assumptions. We are not able to establish a similar result for the case $\Phi_q \equiv 0$. In this case, both the MRC$^w_\phi$, represented by the term $-\Phi_q\Omega'R_w/R_\ell$, and the MHU$_\phi$, represented by $-\Phi_q\Psi'U_c/U_\ell$, disappear so that it is not intuitively obvious whether the two product attributes are greater or smaller than they would be if $\Phi_q > 0$.

Next turn to the long run c-efficient allocation. The long run stock of the durable good, the long run amount of residuals, the long run recyclability of residuals and the long run amount of embodied material are still captured by (5.46). Under the Chichilnisky time preference, however, the discount rate $\delta(t)$ is not constant and positive, but converges to zero as time tends to infinity. Hence, the efficiency conditions are the same as the efficiency conditions (5.47) and (5.48) in the case of the Utilitarian time preference except for replacing $\Psi(\phi) = \int_0^\infty D(a,\phi)e^{-\delta a}da$ by $\Gamma(\phi) = \int_0^\infty D(a,\phi)da$ and $\Omega(\phi) = -\int_0^\infty D_a(a,\phi)e^{-\delta a}da$ by $\Lambda(\phi) = -\int_0^\infty D_a(a,\phi)da$. Since Lemma 2.1 implies $\Lambda = 1$, we immediately obtain

Proposition 5.7. (i) *The stationary state of the c-efficient allocation is characterized by*

$$\frac{1}{Z_\ell} - \frac{Y_m}{Y_\ell} = -\Phi_q\Gamma'\frac{U_c}{U_\ell} + \frac{R_q}{yR_\ell} \,, \tag{5.56}$$

$$\frac{1}{Y_\ell} + q_y\left[\frac{1}{Z_\ell} - \frac{Y_m}{Y_\ell}\right] = -\Gamma\frac{U_c}{U_\ell} + \frac{R_w}{R_\ell} \,. \tag{5.57}$$

(ii) *Suppose $\Phi_q > 0$ and $R_q > 0$ and consider the assumptions*

$$\Gamma, \Phi \text{ are concave: } \Gamma'', \Phi_{qq} < 0 \,, \tag{5.58}$$

$$R \text{ is quasi-linear: } R(\ell_r, w, q_r) = \widetilde{R}(w) + \tilde{\ell}\ell_r + \bar{q}q_r$$
$$\text{with } \widetilde{R}_w > 0, \widetilde{R}_{ww} < 0 \,. \tag{5.59}$$

If (5.49)–(5.51) and (5.58) are satisfied, then the c-efficient long run recyclability and the c-efficient long run durability are greater than they would be if $(\Phi_q > 0, R_q \equiv 0)$ or if $(\Phi_q \equiv 0, R_q \equiv 0)$. If the assumptions (5.49)–(5.51) and (5.59) are satisfied, then the c-efficient long run recyclability and the c-efficient long run durability are greater than they would be if $(\Phi_q \equiv 0, R_q > 0)$.

Proof. Proposition 5.7i follows from (5.47) and (5.48) by replacing Ψ and Ω by Γ and $\Lambda = 1$. To prove Proposition 5.7ii, observe that under (5.49)–(5.51) and (5.58) equation (5.56) simplifies to

$$\widetilde{F}(q_y) := 1/\bar{z} - G(q_y) - \widetilde{H}(q_y) = R_q/yR_\ell \,, \tag{5.60}$$

with $\widetilde{H}(q_y) := \Phi_q\Gamma'\bar{c}/\bar{\ell} > 0$ and $\widetilde{H}_q = \Phi_{qq}\Gamma'\bar{c}/\bar{\ell} + \Phi_q^2\Gamma''\bar{c}/\bar{\ell} < 0$. Hence, $\widetilde{F}$ is increasing in q_y which proves the results for $(\Phi_q > 0, R_q \equiv 0)$ and $(\Phi_q \equiv 0, R_q \equiv 0)$. In order to establish the result for $(\Phi_q \equiv 0, R_q > 0)$, we introduce the shift parameter ϵ and in (5.56) and (5.57) replace Φ_q by $\epsilon\Phi_q$. Hence, $\epsilon = 0$ represents the case $\Phi_q \equiv 0$ and $\epsilon = 1$ stands for the case $\Phi_q > 0$. If the assumptions (5.49)–(5.51) and (5.59) are satisfied, then (5.56) and (5.57) simplify to a system of equations which depends on the variables q_y and $y(=w)$ and on the shift parameter ϵ. Totally differentiating these equations and using Cramer's rule yields

$$\frac{\partial q_y}{\partial \epsilon} = -\frac{1}{|J|} \frac{\bar{c}\Gamma' \Phi_q \tilde{R}_{ww}}{\tilde{\ell}\tilde{\ell}} \ , \qquad \frac{\partial q_y}{\partial \bar{q}} = -\frac{1}{|J|} \frac{\tilde{R}_{ww}}{y\tilde{\ell}^2} \ , \tag{5.61}$$

where $|J|$ is the pertinent Jacobian determinant. From the fact that recyclability for $(\Phi_q > 0, \ R_q > 0)$ is greater than for $(\Phi_q > 0, \ R_q \equiv 0)$ follows $\partial q_y / \partial \bar{q} > 0$ and, hence, the Jacobian determinant has to be positive. Consequently, from (5.61) we obtain $\partial q_y / \partial \epsilon > 0$ which completes the proof of Proposition 5.7ii.   □

Equations (5.56) and (5.57) are the allocation rules for the c-efficient recyclability and the c-efficient production of the durable good in the long run. In contrast to the short run allocation rule (5.44) and (5.45), they contain the same two simplifications described already in the long run u-efficiency scenario. However, there is a further simplification. The MRC_ϕ^w are zero, too. The reason is obvious: For the u-efficient long run durability, this recycling effect is solely due to the positive discount rate whereas in the c-efficient steady state, there is no discounting since the discount rate converges to zero. Therefore, in the c-efficient steady state, the society is indifferent between one unit of recycled material today and one unit of recycled material tomorrow with the consequence that the waste-delaying effect of increasing durability does not cause any recycling benefits or costs.

An immediate consequence of this fundamental difference between long run u-efficiency and long run c-efficiency is revealed by Proposition 5.7ii. The c-efficient long run recyclability and durability are greater for $R_q > 0$ and $\Phi_q > 0$ than they would be if at least one of the derivatives were zero. The result with respect to R_q holds also for the u-efficient steady state whereas the result with respect to Φ_q is valid only in the c-efficient long run. The reason is the following: In case of $\Phi_q > 0$, the c-efficient long run recyclability and durability only yield the additional benefits of increasing the household's utility (MHU_ϕ), but do not cause the additional costs of reducing the recycling productivity via the waste-delaying effect (MRC_ϕ^w) as in the u-efficient steady state. Hence, both product attributes are greater than for $\Phi_q \equiv 0$. This difference between u- and c-efficiency will turn out to be of crucial importance in the subsequent analysis where we evaluate the capacity of markets to implement the efficient allocation.

5.2.2 Benchmark Market System

In this section, we introduce a benchmark market economy with a full set of competitive markets to clarify whether and how the u- and c-efficient allocation can be decentralized by prices. All markets under consideration are assumed to operate smoothly. Prices p_ℓ and p_m for labor and material, respectively, are determined on conventional markets whereas prices for the durable good and residuals are again specified as hedonic price functions. The durable good price $P^y(\phi, q_y)$ depends on the durable's durability and its recyclability, and the residual price $P^w(q_r)$ reflects the household's and recycler's valuation of the residuals' recyclability. So, there are indirect markets for product attributes, i.e. the Walrasian auctioneer determines the price function of the durable good such that demand matches supply of the durable, its durability and its recyclability. The same holds for the residual price which has to equate supply and demand for residuals and their recyclability.

As in Sect. 5.2.1, we investigate both the case in which agents discount future values by a positive and constant rate and the case in which future values are discounted by a variable rate converging to zero. Under the S-rule, the household maximizes the present value of her utility subject to the budget constraint and the integral constraints for the state variables c and w^s and b. Hence, she solves an optimal control problem by maximizing the current-value Lagrangean

$$\mathcal{L}^c(t) = U[c(t), \ell^s(t)] - \int_t^\infty D_a[v - t, \phi^d(t)]y^d(t)q_y^d(t)\lambda_b(v)\rho(v, t)dv$$

$$+ \int_t^\infty D[v - t, \phi^d(t)]y^d(t)\lambda_c(v)\rho(v, t)dv$$

$$- \int_t^\infty D_a[v - t, \phi^d(t)]y^d(t)\lambda_w(v)\rho(v, t)dv$$

$$+ \lambda_\ell(t)\left[p_\ell(t)\ell^s(t) + x(t) + P^w[q_r^s(t)]w^s(t) - P^y[\phi^d(t), q_y^d(t)]y^d(t)\right]$$

$$+ \lambda_{qr}(t)\left[\frac{b(t)}{w^s(t)} - q_r^s(t)\right], \tag{5.62}$$

where $x(t)$ is a lump-sum transfer of profits and net tax revenue to the household at time t. λ_c, λ_w and λ_b are costate variables and all other λs are Lagrange multipliers. The household's budget constraint is represented by the bracketed term in the fourth line of (5.62). It states that the household's expenditures for the durable good must not exceed the household's income consisting of labor income, lump-sum transfers and revenues from selling the consumption residuals to the recycling firm.

The virgin material sector and the recycling sector maximize the present value of their profits subject to their technologies without taking into account any dynamic restriction for state variables. Hence, the dynamic problems of these sectors reduce to a sequence of static problems of maximizing instantaneous profits for all $t \in [0, \infty[$. The pertinent Lagrangeans of the virgin material sector and the recycling sector are, respectively,

$$\mathcal{L}^z(t) = p_m(t)z^s(t) - p_\ell(t)\ell_z^d(t) + \lambda_z(t)\left[Z[\ell_z^d(t)] - z^s(t)\right], \qquad (5.63)$$

$$\mathcal{L}^r(t) = p_m(t)r^s(t) - p_\ell(t)\ell_r^d(t) - P^w[q_r^d(t)]w^d(t)$$

$$+ \lambda_r(t)\left[R[\ell_r^d(t), w^d(t), q_r^d(t)] - r^s(t)\right]. \qquad (5.64)$$

Profits of the virgin material firm in (5.63) equal the difference between revenues from selling the virgin material and labor costs. Profits of the recycling firm in (5.64) equal recycling revenues less labor costs and expenditures of selling the consumption residuals from the household.

It remains to specify the profit maximization problem of the producer. In addition to market prices, we include in the producer's Lagrangean (5.65) a tax rate $\tau_y(t)$ on the supply of the durable good and a tax rate $\tau_m(t)$ on the demand for material at time t. Positive (negative) tax rates represent taxes (subsidies). Both taxes are consistent with the principle of upstream regulation and, hence, possess the advantages discussed in Sect. 5.1.2. Moreover, $\tau_y(t)$ and $\tau_m(t)$ are preferable to direct taxes/subsidies on recyclability and durability since they are based on inputs and outputs which unlike product attributes are relatively easy to measure and to monitor. In view of these convincing advantages, we focus exclusively on the policy (τ_y, τ_m). Analogous to the virgin material and recycling sectors, intertemporal profit maximization of the producer also consists of solving a sequence of static optimization problems since the producer is not confronted with dynamic constraints either. Hence, the pertinent Lagrangean reads

$$\mathcal{L}^y(t) = \left[P^y[\phi^s(t), q_y^s(t)] - \tau_y(t)\right]y^s(t) - p_\ell(t)\ell_y^d(t) - \left[p_m(t) + \tau_m(t)\right]m^d(t)$$

$$+ \lambda_y(t)\left[Y[\ell_y^d(t), m^d(t)] - y^s(t)\right] + \lambda_{qy}(t)\left[\frac{m^d(t)}{y^s(t)} - q_y^s(t)\right]$$

$$+ \lambda_\phi(t)\left[\Phi[q_y^s(t)] - \phi^s(t)\right]. \qquad (5.65)$$

Profits of the durable good producer in (5.65) equal the difference between net revenues from selling the durable good, labor costs and net expenditures for material input. The allocation resulting from (5.62)–(5.65) is said to constitute a *competitive S-equilibrium* if all constraints in (5.31)–(5.41) hold as equalities. Our analysis is restricted to interior solutions.

The Lagrangeans (5.62)–(5.65) yield FOCs which may be rearranged to

$$p_m(t) + \tau_m(t) - p_\ell(t)\frac{Y_m(t)}{Y_\ell(t)} = P_\phi^y(t)\Phi_q(t) + P_q^y(t) , \qquad (5.66)$$

$$\frac{p_\ell(t)}{Y_\ell(t)} + q_y(t)\left[p_m(t) + \tau_m(t) - p_\ell(t)\frac{Y_m(t)}{Y_\ell(t)}\right] = P^y(t) - \tau_y(t) , \quad (5.67)$$

for all $t \in [0,\infty[$, where

$$p_m(t) = \frac{p_\ell(t)}{Z_\ell(t)}, \quad P^w(t) = p_\ell(t)\frac{R_w(t)}{R_\ell(t)}, \quad P_q^w(t) = p_\ell(t)\frac{R_q(t)}{w(t)R_\ell(t)} , \quad (5.68)$$

and

$$P^y(t) = -p_\ell(t)\int_t^\infty D(v)\frac{U_c(v)\psi(v,t)}{U_\ell(v)}dv$$

$$- p_\ell(t)\int_t^\infty D_a(v)\frac{P^w(v)\psi(v,t)}{p_\ell(v)}dv$$

$$+ p_\ell(t)\int_t^\infty D_a(v)\left[q_r(v) - q_y(t)\right]\frac{P_q^w(v)\psi(v,t)}{p_\ell(v)}dv , \qquad (5.69)$$

$$P_\phi^y(t) = -p_\ell(t)\int_t^\infty D_\phi(v)\frac{U_c(v)\psi(v,t)}{U_\ell(v)}dv$$

$$- p_\ell(t)\int_t^\infty D_{a\phi}(v)\frac{P^w(v)\psi(v,t)}{p_\ell(v)}dv$$

$$+ p_\ell(t)\int_t^\infty D_{a\phi}(v)\left[q_r(v) - q_y(t)\right]\frac{P_q^w(v)\psi(v,t)}{p_\ell(v)}dv , \qquad (5.70)$$

$$P_q^y(t) = -p_\ell(t)\int_t^\infty D_a(v)\frac{P_q^w(v)\psi(v,t)}{p_\ell(v)}dv . \qquad (5.71)$$

To check the performance of our benchmark market economy, suppose the government abstains from imposing any taxes and subsidies. Then the capacity of the market economy to achieve the u-efficient or the c-efficient allocation immediately follows from comparing the market equilibrium conditions (5.66)–(5.71) with the efficiency conditions (5.44) and (5.45).

Proposition 5.8. *For all $t \in [0,\infty[$, set $\tau_y(t) = \tau_m(t) = 0$, $p_\ell(t) = -U_\ell(t)$ and the remaining prices as in (5.68)–(5.71). Then, the competitive S-equilibrium is u-efficient in case of the Utilitarian time preference and c-efficient in case of the Chichilnisky time preference.*

Proposition 5.8 states that our benchmark economy is capable of supporting the u-efficient allocation in case of the Utilitarian time preference and the c-efficient allocation in case of the Chichilnisky time preference. Moreover, Proposition 5.8 together with (5.68)–(5.71) provides information about how prices bring about the u-efficient allocation and the c-efficient allocation, respectively. The price for labor, $p_\ell(t)$, captures the marginal disutility of labor. The price for material, $p_m(t)$, reflects the marginal extraction costs of primary material. The price for residuals, $P^w(t)$, and the indirect price for the residuals' recyclability, $P_q^w(t)$, equal marginal recycling benefits of residuals and marginal recycling benefits of the residuals' recyclability, respectively. The price for the durable good at time t, $P^y(t)$, comprises the MHU_y, the MRB_y^w and the MRB_y^q. The indirect price for durability at time t, $P_\phi^y(t)$, contains the MHU_ϕ, the MRC_ϕ^w and the MRB_ϕ^q. The price for the durable's recyclability at time t, $P_q^y(t)$, reflects the MRB_q^q. In summary, all marginal costs and benefits of the various variables are signaled by market prices and, hence, the competitive S-equilibrium decentralizes an u-efficient and c-efficient allocation, respectively.

In general, it is not possible to specify the sign of all equilibrium prices. While the prices for labor, material, residuals and recyclability of residuals are positive at all times, the prices for the durable good, its durability and its recyclability are ambiguous in sign. Focusing on the long run yields more clear-cut results. The u-efficient prices in Proposition 5.8 then simplify to

$$p_\ell = -U_\ell > 0\,, \quad p_m = \frac{p_\ell}{Z_\ell} > 0\,, \tag{5.72}$$

$$P^w = p_\ell \frac{R_w}{R_\ell} > 0\,, \quad P_q^w = p_\ell \frac{R_q}{yR_\ell} > 0\,, \tag{5.73}$$

$$P^y = -p_\ell \frac{U_c}{U_\ell}\Psi + P^w\Omega > 0\,, \quad P_\phi^y = -p_\ell \frac{U_c}{U_\ell}\Psi' + P^w\Omega' \gtreqless 0\,, \tag{5.74}$$

$$P_q^y = P_q^w\,\Omega > 0\,. \tag{5.75}$$

Hence, in our benchmark economy, the u-efficient long run prices are all unambiguously positive except for the price of durability. The latter may be negative since the MHU_ϕ, represented by $-p_\ell U_c\Psi'/U_\ell > 0$, and the MRC_ϕ^w, represented by $P^w\Omega' < 0$, are opposite in sign. Hence, if the absolute value of the MRC_ϕ^w is greater than the MHU_ϕ, then the price of durability is negative since the recycling costs of durability overcompensate the consumption benefits of durability. In contrast, the c-efficient long run prices are given by (5.72), (5.73) and

$$P^y = -p_\ell \frac{U_c}{U_\ell}\Gamma + P^w > 0\,, \quad P_\phi^y = -p_\ell \frac{U_c}{U_\ell}\Gamma' > 0\,, \quad P_q^y = P_q^w > 0\,. \tag{5.76}$$

Thus, the c-efficient price for durability is now unambiguously positive. The reason is that in the long run c-efficiency scenario, the MRC_ϕ^w disappear and, hence, the price of durability reflects the MHU_ϕ only.

5.2.3 Market Failure and its Correction

For the design of an effective policy, it is crucial to assess the empirical relevance of the market economy discussed above. As in Sect. 5.1, it is now
assumed that markets for product characteristics are absent. In doing so, we
proceed stepwise starting with the case of missing markets for recyclability,
then turning to the case without price signals for durability and finally investigating the case in which markets for both recyclability and durability are
missing.

No Indirect Markets for Recyclability

If markets for recyclability are absent, then there are no prices available that
depend on recyclability, i.e. $P_q^y(t) = P_q^w(t) = 0$ for all $t \in [0,\infty[$. Without
any corrective policy the consequences are (a) that the production firm determines the level of recyclability without any regard of the household's and
the recycler's needs or wants and (b) that the household and the recycler
take recyclability as given. Recyclability then represents an externality for
the household and the recycling firm. For this scenario, we obtain[14]

Proposition 5.9. *Suppose there are no markets for recyclability, i.e.* $P_q^y(t) = P_q^w(t) = 0.$
(i) Set $\tau_y(t) = \tau_m(t) = 0$ *for all* $t \in [0,\infty[$. *Then, in case of the Utili-*
tarian time preference, the competitive S-equilibrium is characterized by an
u-inefficient recyclability and durability for all $t \in [0,\infty[$, *in general. In case*
of the Chichilnisky time preference, the competitive S-equilibrium is charac-
terized by a c-inefficient recyclability and durability for all $t \in [0,\infty[$, *in gen-*
eral.[15]
(ii) Set all remaining prices as in Proposition 5.8 and set

$$\tau_m(t) = -U_\ell(t) \int_t^\infty \left[D_a(v) + \Phi_q(t) D_{a\phi}(v) \Big[q_y(t) - q_r(v) \Big] \right] \frac{R_q(v)\psi(v,t)}{w(v)R_\ell(v)} dv \, ,$$

$$\tau_y(t) = U_\ell(t) \int_t^\infty \Bigg[q_r(v) D_a(v)$$

$$- \Phi_q(t) q_y(t) D_{a\phi}(v) \Big[q_r(v) - q_y(t) \Big] \Bigg] \frac{R_q(v)\psi(v,t)}{w(v)R_\ell(v)} dv \, ,$$

---

[14] To prove Proposition 5.9, set $P_q^y(t) = P_q^w(t) = 0$ in (5.66)–(5.71) and compare
the resulting expressions with (5.44) and (5.45).

[15] The phrase 'in general' means that it can not be excluded that the product characteristics are incidentally efficient. But this will be the case only under some very
special circumstances under which the opposing distortions incidentally cancel
out.

for all $t \in [0, \infty[$. Then, the competitive S-equilibrium is u-efficient in case of the Utilitarian time preference and c-efficient in case of the Chichilnisky time preference.

The reason for market failure identified in Proposition 5.9i is that in absence of markets for recyclability, there are no prices which capture the MRB_q^q, the MRB_ϕ^q and the MRB_y^q. The producer of the durable good chooses the recyclability inappropriately low or high because she does not receive any price signals regarding the productivity of recyclability in the recycling process. Since recyclability is positively correlated with durability, the latter product attribute is also set inappropriately. Placing these results into the context of the existing literature two remarks are in order. First, *u-inefficiency* with respect to recyclability has already been demonstrated by Eichner and Pethig (2001) in a static general equilibrium framework of a nondurable consumption good. In contrast, the *u-inefficiency* with respect to durability represents an extension of the durability literature: As is obvious from the articles already referred to in the introduction and from the analysis in the previous chapters, the extant literature identifies imperfect competition and environmental externalities as reasons for an u-inefficient provision of product durability in unregulated markets. Proposition 5.9i reveals that durability may be u-inefficient even in a perfectly competitive market in which environmental externalities are absent. In our model, durability is u-inefficient since it is related to recyclability which, in turn, is distorted by the recyclability externality. Second, the previous literature and the models of the previous chapters do not provide information about the capability of markets to generate the *c-efficient* product design. But this is done by Proposition 5.9i: In case of missing markets for recyclability, the unregulated S-equilibrium does not provide c-efficient levels of product attributes.

Proposition 5.9ii specifies a tax-subsidy scheme which is capable of correcting market failure identified in Proposition 5.9i. This policy option influences the producer's product design in an indirect way since it suggests to regulate output of the durable good together with material input rather than the product attributes recyclability and durability directly. The instruments contain the MRB_q^q, the MRB_ϕ^q and the MRB_y^q which in the benchmark economy are reflected by competitive prices, but in the present scenario, do not emerge in the price system since markets for recyclability are absent. In general, it is not possible to determine the signs of the tax rates, but more specific results are obtained if we pay attention to the the stationary state. Let us start with the case of the Utilitarian time preference.[16]

---

[16] Proposition 5.10 is easily proved by comparing the pertinent conditions for the market equilibrium and the efficient allocation and by observing that the long run laissez faire allocation for $P_q^y = P_q^w = 0$ equals the efficient allocation for $R_q \equiv 0$ which, in case of the Utilitarian time preference, is characterized by Proposition 5.6ii.

Proposition 5.10. *Suppose agents have Utilitarian time preference and there are no markets for recyclability, i.e.* $P_q^y(t) = P_q^w(t) = 0$.
(i) If (5.49)–(5.54) are satisfied and τ_y *and* τ_m *are set equal to zero, then the steady state of the competitive S-equilibrium is characterized by an* <u>*u-inefficiently low*</u> *recyclability and an* <u>*u-inefficiently low*</u> *durability.*
(ii) Set all remaining prices as in Proposition 5.8 and set

$$\tau_y = -\tau_m q_y = -q_y \Omega \frac{R_q}{yR_\ell} U_\ell > 0 \, .$$

Then, the long run competitive S-equilibrium is <u>*u-efficient.*</u>

In the long run, the durable's recyclability and the residuals' recyclability are the same. The MRB_ϕ^q and the MRB_y^q then are zero and the MRB_q^q is the only effect which is not represented by prices when markets for recyclability are absent. Based on these simplifications, Proposition 5.10 specifies the result of Proposition 5.9 under the Utilitarian time preference in two aspects. First, the unregulated S-equilibrium in the long run generates u-inefficiently *low* product attributes if markets for recyclability break down. Although the underlying assumptions (5.49)–(5.54) are quite restrictive, we conjecture that the result is valid under less restrictive conditions, too, since the ignored MRB_q^q represented by $\Omega R_q/yR_\ell$ are unambiguously positive. Second, the signs of the tax rates capable of restoring u-efficiency in the long run are unambiguous. The long run regulatory scheme in Proposition 5.10ii contains an output tax $(\tau_y > 0)$ and a material subsidy $(\tau_m < 0)$. The output tax induces the producer to reduce her supply of the good and the material subsidy stimulates her demand for material input. The net effect of both instruments is to shift recyclability up to its u-efficient long run level and, thus, long run durability becomes u-efficient, too. In the recycling literature, this policy option is known as deposit-refund system (Palmer and Walls, 1997).

The steady state in case of the Chichilnisky time preference differs from the steady state in case of the Utilitarian time preference since the MRC_ϕ^w disappear, too, and since the functions Ψ and Ω are replaced by Γ and $\Lambda = 1$, respectively. But these changes do not alter the result of Proposition 5.10 in a qualitative way when we switch from the Utilitarian to the Chichilnisky time preference. The only difference is the magnitude of the tax rate since the function Ω takes on the value 1. Consequently, to achieve c-efficiency in the steady state of the competitive S-equilibrium, a deposit-refund system needs to be implemented to overcome the c-inefficiently low recyclability and the c-inefficiently low durability. A qualitative difference between u-efficiency restoring policies and c-efficiency restoring policies is derived in the next section.

No Indirect Markets for Durability

Suppose now that indirect markets for recyclability are working smoothly, but markets fail to provide price signals for durability. As a consequence, the price functions depend on recyclability, but not on durability, i.e. $P^y_\phi(t) = 0$ for all $t \in [0, \infty[$. In this scenario, three marginal effects are not captured by market prices: the MHU_ϕ, the MRC^w_ϕ and the MRB^q_ϕ. Analogous to the analysis of missing markets for recyclability in Proposition 5.9, the unregulated S-equilibrium can then be shown to be neither u-efficient nor c-efficient, in general. In contrast to the previous paragraph, however, now the market failure no longer is due to the recyclability externality, but due to the durability externality: Since markets for durability fail, the producer does not obtain correct price signals for durability and, hence, she chooses an inappropriate level for this product attribute as well as for the closely related recyclability.

It can be shown that the signs of the tax-subsidy rates restoring u- and c-efficiency are ambiguous, in general. More clear-cut results are again obtained if we restrict the focus on the stationary state as shown in the following two propositions.[17]

Proposition 5.11. *Suppose agents have Utilitarian time preference and there are no markets for durability, i.e. $P^y_\phi(t) = 0$.*
(i) If τ_y and τ_m are set equal to zero, then recyclability and durability in the steady state of the competitive S-equilibrium may deviate from their u-efficient levels in either direction even under the restrictive assumptions (5.49)–(5.54).
(ii) Set all remaining prices as in Proposition 5.8 and set

$$\tau_y = -\tau_m q_y = q_y \Phi_q \Psi' U_c - q_y \Phi_q \Omega' \frac{R_w}{R_\ell} U_\ell \gtreqless 0 \ .$$

Then, the long run competitive S-equilibrium is u-efficient.

Proposition 5.12. *Suppose agents have Chichilnisky time preference and there are no markets for durability, i.e. $P^y_\phi(t) = 0$.*
(i) If (5.49)–(5.51) and (5.59) are satisfied and τ_y and τ_m are set equal to zero, then the steady state of the competitive S-equilibrium is characterized by a c-inefficiently low recyclability and a c-inefficiently low durability.
(ii) Set all remaining prices as in Proposition 5.8 and set

$$\tau_y = -\tau_m q_y = q_y \Phi_q \Gamma' U_c > 0 \ .$$

Then, the long run competitive S-equilibrium is c-efficient.

If the time preference is Utilitarian and if markets for durability are absent, then there are still *two* marginal effects of product design not captured by prices in the long run S-equilibrium: the MHU_ϕ and the MRC^w_ϕ. These effects are opposite in sign and, consequently, Proposition 5.11i tells us that it is not clear whether recyclability and durability are u-inefficiently low or high. It follows that the signs of the u-efficiency restoring tax-subsidy rates in Proposition 5.11ii are indeterminate even in the steady state. Hence, in this

---

[17] These propositions are proved in the same way as Propositions 5.9 and 5.10.

case, the deposit-refund system is not necessarily an appropriate policy – a result which is opposed to the result derived in Proposition 5.10 for the case of absent markets for recyclability. In contrast, if we consider the Chichilnisky time preference and if there are no markets for durability, then in the long run S-equilibrium only the MHU_ϕ are not reflected by the price system while the MRC_ϕ^w are identical to zero. Consequently, Proposition 5.12 shows that the unregulated product attributes are c-inefficiently *low* and that c-efficiency is restored by implementing a deposit-refund system, i.e. by subsidizing material input and taxing output of the production sector. This result is qualitative the same as in case of missing markets for recyclability.

So far, our analysis was based on the S-rule. We now wish to elaborate the consequences of changing the property rights regime from the S-rule to the R-rule. Under the R-rule, the producer rents the durable good and sells the services (rather than the good itself) at price p_c while the durables remain her property. Moreover, the producer is the owner of residuals since at the end of the product's life the household hands over the goods to the producer who sells residuals at price $P^w(q_r^s)$ to the recycler.[18] It is straightforward to show that the efficient allocation under the R-rule is the same as under the S-rule. This is in accordance with Coasean economics with zero regime specific transaction costs. Hence, (5.44) and (5.45) are the efficiency conditions not only under the S-rule, but also under the R-rule.

To derive the conditions characterizing a competitive equilibrium under the R-rule, note that the Lagrangeans (5.63) and (5.64) of the extraction firm and the recycling firm remain unchanged. However, state variables are now taken into account by the producer since she owns the stock of the durables and supplies the consumption residuals to the recycling firm. Consequently, profit maximization of the durable good firm is an optimal control problem with state constraints for c^s, w^s and b. The pertinent current-value Lagrangean reads

$$\mathcal{L}^y(t) = \left[p_c(t) - \tau_c(t)\right]c^s(t) + P^w[q_r^s(t)]w^s(t) - p_\ell(t)\ell_y^d(t)$$

$$- \left[p_m(t) + \tau_m(t)\right]m^d(t) + \int_t^\infty D[v - t, \phi(t)]y(t)\lambda_c(v)\rho(v,t)dv$$

$$- \int_t^\infty D_a[v - t, \phi(t)]y(t)\lambda_w(v)\rho(v,t)dv$$

$$- \int_t^\infty D_a[v - t, \phi(t)]y(t)q_y(t)\lambda_b(v)\rho(v,t)dv$$

---

[18] Note that under the R-rule there exists no price for the durable good ($P^y = P_q^y = P_\phi^y = 0$).

$$+ \lambda_y(t) \left[Y[\ell_y^d(t), m^d(t)] - y(t) \right] + \lambda_{qv}(t) \left[\frac{m^d(t)}{y(t)} - q_y(t) \right]$$

$$+ \lambda_\phi(t) \left[\Phi[q_y(t)] - \phi(t) \right] + \lambda_{qr}(t) \left[\frac{b(t)}{w^s(t)} - q_r^s(t) \right] , \qquad (5.77)$$

where $\tau_c(t)$ is the tax rate levied on the durable stock at time t. The first four terms in (5.77) yield the producer's profits which equal the difference between revenues and costs. Revenues stem from renting the durable good to the household and from selling the residuals to the recycler. Costs consist of labor and material costs. The representative household does not take into account state variables and, hence, her problem of maximizing the present value of utility is equivalent to a sequence of static problems maximizing instantaneous utility for all $t \in [0, \infty[$. The pertinent Lagrangean reads

$$\mathcal{L}^c(t) = U[\ell^s(t), c^d(t)] + \mu_\ell(t) \left[p_\ell(t)\ell^s(t) + x(t) - p_c(t)c^d(t) \right] , \qquad (5.78)$$

where the bracketed term represents the household's budget constraint which states that expenditures for renting the durable must not exceed the sum of labor income and lump-sum transfers. The allocation resulting from solving (5.63), (5.64), (5.77) and (5.78) is said to constitute a *competitive R-equilibrium* if all constraints in (5.31)–(5.41) hold as equalities. Again, only interior solutions are of interest.

The FOCs related to (5.63), (5.64), (5.77) and (5.78) can be rearranged to yield

$$\frac{p_m(t) + \tau_m(t)}{p_\ell(t)} - \frac{Y_m(t)}{Y_\ell(t)} + \Phi_q(t) \int_t^\infty D_{a\phi}(v) \frac{P^w(v)\rho(v,t)}{p_\ell(t)} dv =$$

$$\Phi_q(t) \int_t^\infty D_\phi(v) \frac{[p_c(v) - \tau_c(v)]\rho(v,t)}{p_\ell(t)} dv - \int_t^\infty D_a(v) \frac{P_q^w(v)\rho(v,t)}{p_\ell(t)} dv$$

$$- \Phi_q(t) \int_t^\infty D_{a\phi}(v) \left[q_y(t) - q_r(v) \right] \frac{P_q^w(v)\rho(v,t)}{p_\ell(t)} dv , \qquad (5.79)$$

and

$$\frac{1}{Y_\ell(t)} + q_y(t) \left[\frac{p_m(t) + \tau_m(t)}{Z_\ell(t)} - \frac{Y_m(t)}{Y_\ell(t)} \right] =$$

$$\int_t^\infty D(v) \frac{[p_c(v) - \tau_c(v)]\rho(v,t)}{p_\ell(t)} dv - \int_t^\infty D_a(v) \frac{P^w(v)\rho(v,t)}{p_\ell(t)} dv$$

$$+ \int_t^\infty D_a(v) \left[q_r(v) - q_y(t) \right] \frac{P_q^w(v)\rho(v,t)}{p_\ell(t)} dv , \qquad (5.80)$$

for all $t \in [0, \infty[$ where

$$p_m(t) = \frac{p_\ell(t)}{Z_\ell(t)}, \qquad p_c(t) = -p_\ell(t)\frac{U_c(t)}{U_\ell(t)}, \qquad (5.81)$$

$$P^w(t) = p_\ell(t)\frac{R_w(t)}{R_\ell(t)}, \qquad P_q^w(t) = p_\ell(t)\frac{R_q(t)}{yR_\ell(t)}. \qquad (5.82)$$

Comparing (5.79)–(5.82) with the efficiency conditions (5.44) and (5.45) immediately yields

Proposition 5.13. *Suppose the durable is rented. For all $t \in [0, \infty[$, set $\tau_c(t) = \tau_m(t) = 0$, $p_\ell(t) = -U_\ell(t)$ and the other prices as in (5.81) and (5.82). Then the competitive R-equilibrium is u-efficient in case of the Utilitarian time preference and c-efficient in case of the Chichilnisky time preference.*

The unregulated competitive R-equilibrium with rental markets for durable goods provides an u-efficient and a c-efficient allocation, respectively. Comparing the S-rule under failing indirect markets for durability (Propositions 5.11 and 5.12) and the R-rule in the same scenario (Proposition 5.13) offers an important insight for policy options: The market failure due to missing markets for durability may be corrected either by means of a tax-subsidy scheme or, alternatively, through the introduction of the R-rule under which the producer rents (voluntarily or through regulatory order) rather than sells her goods. The reason for this result is straightforward: Reassigning the property rights for the consumption goods from the household to the producer internalizes the durability externality since the 'durable stock production' represented by (5.35), (5.37) and (5.38) is integrated into the production sector. Hence, the producer takes into account the marginal effects of durability on the durable stock and on residuals.

No Indirect Markets for Recyclability and Durability

After having investigated the consequences of missing markets for recyclability and durability separately, we now analyze both cases simultaneously by assuming that the price functions P^y and P^w neither depend on recyclability nor on durability. Therefore, the market economy is now distorted by a recyclability externality and a durability externality. Not surprisingly, for this scenario, it can be shown that under the S-rule the same qualitative results hold as in the previous section since the recyclability externality is merely added to the durability externality of the previous section. The only difference to Proposition 5.11 and 5.12 is that the tax-subsidy rates now also take care of the MRB_q^q. Hence, the unregulated S-equilibrium is shown to provide an inappropriate product design and the tax-subsidy rates which are capable of correcting the market failure are ambiguous in sign except for the steady state under the Chichilnisky time preference.

Finally, we address the question whether the R-rule is also a suitable measure to cope with a break down of indirect markets for both recyclability and durability. For the stationary state, this question is answered with the help of[19]

Proposition 5.14. *Suppose agents have Utilitarian time preference and the durable consumption good is rented. Furthermore, assume there are no markets for recyclability and durability, i.e.* $P_q^y(t) = P_q^w(t) = P_\phi^y(t) = 0$.
(i) If (5.49)–(5.54) are satisfied and τ_c *and* τ_m *are set equal to zero, then the steady state of the competitive R-equilibrium is characterized by an* u-inefficiently low *recyclability and an* u-inefficiently low *durability.*
(ii) Set all remaining prices as in Proposition 5.13 and set

$$\tau_c \Phi_q \Psi = -\tau_m q_y = -\frac{q_y \Omega \Psi}{\Psi - q_y \Phi_q \Psi'} \frac{R_q}{y R_\ell} U_\ell > 0 .$$

Then, the long run competitive R-equilibrium is u-efficient.

For the case of a variable discount rate, Proposition 5.14 carries over except that the c-efficiency restoring tax-subsidy scheme reads

$$\tau_c \Phi_q \Gamma = -\tau_m q_y = -\frac{q_y \Gamma}{\Gamma - q_y \Phi_q \Gamma'} \frac{R_q}{y R_\ell} U_\ell > 0 . \tag{5.83}$$

These results show that adopting the R-rule alone is not sufficient to provide the appropriate allocation in case of missing price signals for recyclability and durability. An additional tax-subsidy scheme (τ_c, τ_m) is needed which consists of a material subsidy and a tax on the rented durables to capture the marginal recycling benefits of product design. In view of the functional relation between durability and recyclability, Proposition 5.14 has to be considered a surprise. The explanation of this proposition, however, is straightforward: If the producer rents the products, then she keeps the property rights for both the consumption goods and the residuals and therefore it is in her own interest to take into account how durability influences the stock of the durable good and the residuals' recyclability. Thus, as in Proposition 5.13, the durable stock accumulation is integrated in the production sector and the durability externality is internalized. But shifting the property rights of residuals from households to producers has no impact on the recyclability externality since the R-rule does not integrate the recycling sector into the production sector. As a consequence, the recyclability externality has to be internalized through a tax-subsidy scheme or, alternatively, by vertically integrating the recycling firm into the production sector.

---

[19] To prove the sign of the tax-subsidy rates, note that owing to Ψ'', $\Phi_{qq} \leq 0$ the function Ψ is concave in q and hence we obtain $\Psi/q > \Psi' \Phi_q$. The same observation holds for Γ.

5.3 Summary and Practical Considerations

The main objective of this chapter was to explicitly account for recycling in the analysis of durable consumption goods. For this purpose, Sect. 5.1 has investigated a durable good economy with fixed product durability and endogenous recyclability which positively affects the productivity in the recycling sector. While contributions to the literature concluded that a subsidy on recyclability is sufficient to restore efficiency when markets for recyclability are absent, our analysis reveals that this is true only in economies with nondurable consumption goods (as is implicitly presupposed in previous studies) or in the steady state of economies with durable consumption goods as shown in equation (5.30). According to Proposition 5.4, however, the efficient policy on the adjustment path of a durable good economy comprises an additional tax or subsidy on the output of the consumption good since this output creates an externality in the recycling sector, too. Furthermore, we gain the insight that on the adjustment path, the subsidy on recyclability may turn into a tax since the policy instrument on recyclability is indeterminate in sign. So, the analysis in Sect. 5.1 shows that there are significant differences between regulation of nondurable goods or long run regulation of durable goods, on the one hand, and short run regulation of durable goods, on the other hand.

From a practical point of view, short run aspects of green design are relevant in present durable good markets. For example, the data of the German Association of Automobile Manufactures suggest that the number of new automobiles sold in Germany varied considerably over the last decades (VDA, 2001). It boosted from 1 million cars in 1960 to 4.1 million cars in 1991 and decreased thereafter to 3.3 million cars in 2000 with a 11.1 (!) percent reduction from 1999 to 2000. Moreover, the figures presented in the introduction to this chapter show that the recyclable share of each car has been increased in the past and further improvements in product recyclability are to be expected. These observations indicate that important durable good markets have not yet reached a steady state (and perhaps never will), but are still on an adjustment path. A short run perspective is therefore important, and policy makers ought to pay special attention to the short run recycling policy of durable consumption goods.

The analysis in Sect. 5.2 extends the model of Sect. 5.1 by making durability an explicit choice variables of producers and by postulating a functional relation between product durability and recyclability. Moreover, it accounts for aspects of intergenerational distribution by considering, alternatively, the Utilitarian time preference (u-efficiency) and the Chichilnisky time preference (c-efficiency). As in Sect. 5.1, the main objective was to characterize market failure and to offer policy recommendations in case markets for product attributes fail to be active. The findings of this analysis may be summarized as follows:

1. If indirect markets for at least one of the product attributes are absent, then both recyclability and durability of the consumption goods are u-

and c-inefficient, respectively (Propositions 5.9i, 5.10i, 5.11i, 5.12i). Hence, with regard to durability, our analysis shows that besides imperfect competition and environmental externalities investigated in the previous chapters there is a further source of distortions in the allocation of durability, namely missing markets for product design.

2. If markets for recyclability are absent and if the economy has reached its steady state, then a deposit-refund system, i.e. a tax on output and a subsidy on material input of the durable good producer, is capable of restoring u- and c-efficiency, respectively (Proposition 5.10ii).

3. If markets for durability are absent, then the signs of the u-efficiency restoring tax and subsidy rates are ambiguous even in the steady state and, hence, in this case the deposit-refund system is not necessary an appropriate policy (Proposition 5.11ii). In contrast, with the Chichilnisky time preference, the deposit-refund system remains a suitable instrument to restore long run c-efficiency (Proposition 5.12ii). Consequently, our analysis reveals a fundamental difference between regulation under the Utilitarian time preference and regulation under the Chichilnisky time preference and shows that the specific form of time preference is crucial for efficient recycling policy of consumer durables.

4. Focusing on property rights reveals another promising policy recommendation: In case of absent markets of durability, inducing producers to rent rather than to sell durables turns out to internalize the durability externality independent of the type of time preference and independent of whether the economy is in its steady state (Proposition 5.13).

5. The efficiency-enhancing capacity of changing property rights is limited, however, since in case that both indirect markets for product attributes are absent an additional deposit-refund system is needed that reflects marginal recycling benefits of recyclability (Proposition 5.14).

We will finally discuss some practical aspects of property rights regulation summarized in points 4 and 5 above. In case of missing markets for product durability, the obligation to rent the durable good proposed in Proposition 5.13 has an important advantage over the tax-subsidy schemes suggested in Propositions 5.11ii and 5.12ii. Information requirements for implementing the optimal tax-subsidy schemes are quite demanding. The policy maker needs to assess the marginal utility of the durable good, the marginal recycling productivity of residuals and of their recyclability, the decay function and the relationship between recyclability and durability of the consumption good. Moreover, she needs to find out whether the economy already reached its steady state or is still on the adjustment path and whether individuals have Utilitarian or Chichilnisky time preference. On the other hand, the policy option of reassigning property rights does not require *any* of this information. The policy maker simply mandates the producer to rent the durable good and then market forces establish the efficient allocation. Hence, we conclude that in case of missing price signals for durability, the reassignment of property rights seems to be a powerful instrument to overcome market failure.

In case that markets for durability and recyclability break down simultaneously, inducing firms to rent does not suffice to attain efficiency since, in addition, we need a tax-subsidy scheme according to Proposition 5.14. Regardless of this restriction in the effectiveness of the R-rule, it is worth mentioning that the policy of changing property rights still has an advantage with respect to information requirements. The tax-subsidy scheme which is necessary to overcome market failure under the S-rule requires to assess the marginal utility of durability, the marginal recycling productivity of residuals and their recyclability, the decay function and the relation between recyclability and durability. In contrast, the tax-subsidy scheme which is necessary to overcome market failure under the R-rule does not require information about the marginal utility and the marginal recycling productivity. Hence, with respect to information requirements, the reassignment of property rights is still superior to the tax-subsidy scheme under the S-rule.

While this advantage of property rights regulation is appealing, two remarks are in order which strain the favorable image of the R-rule. First, our formal analysis has shown that inducing producers to rent their products internalizes the durability externality. The hope of the advocates of the R-rule is that renting internalizes also the environmental externalities raised by the durable good (e.g. Soete, 1997). However, our formal model completely ignores environmental costs and it cannot be expected that the R-rule internalizes such costs if they were explicitly taken into account. The reason is that the R-rule does not integrate the 'environmental sector' into the production sector. Hence, property rights regulation is capable of correcting market failure caused by missing price signals for product characteristics, indeed, but is not appropriate to correct market failure based on environmental externalities. Such externalities still require other instruments as, for example, a Pigouvian waste tax.

Second, renting is often entailed with the problem that the owner and the user of the product do not coincide. The producer (or a firm instructed by the producer) is the owner of a rented durable while the consumer is the user. Of course, under this property rights regime, the producer prefers frequent maintenance and a thorough treatment of the product. But the consumer usually does not possess incentives for such activities since once the durable is out of order she may rent another unit of the durable good. Hence, in rental markets, there is a negative effect on product durability which our formal model in Sect. 5.2 does not take into account, but which tends to weaken the efficiency capacity of the R-rule. Another problem of the R-rule also based on the divergence between owner and user is that many consumers see the durable good as status symbol. Important examples are cars or mobile phones. Hence, inducing consumers to rent the products brings about a welfare loss since consumers derive utility not only from the services of the durable good, but also from the property rights at the product. This welfare loss is ignored in our formal model, but again weakens the efficiency capacity of the R-rule.

6

Conclusion

In this concluding chapter, we will first summarize the answers which our analysis provides to the question raised in the introduction and then briefly discuss further lines of research in environmental policy for consumer durables. Let us start with summarizing the main results of our analysis.

(1) *Allocative Efficiency*

The first question we raised in the introduction was how the efficient allocation in durable good economies with environmental pollution or recycling is characterized. Roughly speaking, efficiency is attained if all variables are set such that the pertinent marginal social costs equal the pertinent marginal social benefits. Of special interest in our context is the impact of increasing product durability on social costs and benefits. In all settings considered in the present study, an increase in durability raises production costs (since it is more expensive to render the products more durable) and consumer benefits (since the stock of the durable good is enhanced).

Besides these two general effects, raising durability also influences environmental, disposal and recycling costs. The environmental costs caused by production emissions of the durable good decline due to the emissions-reducing effect considered in the model of Chap. 3. The same is true for the disposal costs of the durable good considered in Chap. 4 if we focus on the long run. In such a setting, both the waste-delaying and the waste-reducing effect of durability curtail disposal costs. However, on the adjustment path, the waste-delaying effect may *increase* disposal costs, namely if marginal disposal costs at early dates are sufficiently lower than at later dates. In the durable good model with recycling presented in Chap. 5, increasing durability has an impact on social costs and benefits through the waste-delaying effect, too, i.e. it reduces the benefits from recycling at early dates (since the amount of residuals and the recycling output decline at these dates) and raises the benefits of recycling at later dates (since the amount of residuals and the recycling output increase in later periods). In the long run, the net effect is an increase in recycling costs if time preference is Utilitarian since the society then prefers early recycling output due to the positive discount rate. In contrast, under Chichilnisky time preference, the net effect of the waste-delaying effect disappears in the long run since society is then indifferent between a unit of recycled material today and a unit of recycled material tomorrow.

Since these effects of product durability partly point into different directions, an increase in product durability is neither per se socially beneficial nor per se beneficial from an environmental point of view. Of course, whether improved durability of consumer durables raises or reduces overall social and environmental costs is an empirical question. Answering this question is beyond the scope of the present study and, therefore, it is an issue for future empirical research on environmental policy in markets for consumer durables.

(2) Market Equilibrium and Laissez Faire

The analysis in the previous chapters have shown that the presence of environmental, disposal or recycling costs leads to market failure in a laissez faire durable good economy. With respect to durability, we have derived both overprovision and underprovision results. Laissez faire product durability is unambiguously smaller than efficient product durability, for example, (a) in durable good economies with production emissions since unregulated producers ignore the emissions-reducing effect (Chap. 3), (b) in the stationary state of durable good economies with disposal costs of solid waste and commitment ability of producers since unregulated producers do not take into account the reduction in disposal costs due to the waste-delaying and the waste-reducing effects (Chap. 4) and (c) in the stationary state of durable good economies with recycling and missing markets for product design since producers ignore the positive effect of their product design on durable goods recycling (Chap. 5).

Nevertheless, our analysis identifies three major reasons for an overprovision of product durability. First, as already argued above, the waste-delaying effect may cause disposal costs on the adjustment path (Chap. 4). In that case, increasing durability exerts a negative externality. This externality is not taken into account by unregulated producers with the consequence that laissez faire durability may be inefficiently high. Second, imperfectly competitive durable good sellers without commitment ability have an incentive for planned obsolescence which tends to render durability inefficiently low and an incentive for revenue raising which tends to shift durability beyond its efficient level (Sect. 4.1). If the latter effect is strong enough to overcompensate the former effect and the environmental externality, then an overprovision of laissez faire durability occurs. Third, in durable good economies with recycling, product durability has various countervailing effects on social costs and benefits (Chap. 5). It influences the recycling output through the waste-delaying effect or through changes in the product recyclability. In general, it is therefore not clear whether durability causes a positive or negative externality and, hence, it cannot be excluded that product durability in a laissez faire economy is greater than in the social optimum. Of course, whether durability in real world markets is inefficiently low or high is again an empirical question.

(3) *First-Best Regulation*

Concerning regulatory policies capable of restoring efficiency, one basic result of our analysis is that in many cases, efficiency in polluting durable good economies is attained by the same instruments as in polluting nondurable good economies. Under perfect competition, a Pigouvian tax on production emissions (Chap. 3) or on solid waste (Chap. 4) is sufficient to correct for market failure since this tax completely internalizes the environmental externality. Under imperfect competition, the Pigouvian tax on emissions (Chap. 3) or on waste (Chap. 4) has to be complemented by a subsidy on the stock of the durable such that the market imperfection due to the market power of durable good producers is removed, too. Both policy approaches are already well known from environmental regulation of polluting nondurable good economies.

Nevertheless, our analysis suggests that there are cases in which regulation of polluting durable good producers significantly differs from regulation of polluting nondurable good producers. Two cases are of special interest. First, if durable good sellers act under imperfect competition and possess no commitment ability, the efficiency restoring regulatory scheme has to account for the firms' incentives for planned obsolescence and revenue raising (Sect. 4.2). So, the regulator either needs to levy an additional tax on product durability or has to adjust the tax on waste or the subsidy on the stock of the durable good such that these instruments account for the additional distortions of product durability. Second, regulation for durable goods and nondurable goods significantly differs also in case that recycling of the consumption good is influenced by the product's recyclability, the markets for this recyclability fail to be active and the economy under consideration has not yet reached the stationary state, but is still on its adjustment path (Sect. 5.1). While for nondurable goods a Pigouvian subsidy on recyclability suffices to restore efficiency in this case, for durable goods, we have to employ an additional output tax or subsidy since the output of the durable good generates an externality in the recycling sector, too.

(4) *Second-Best Regulation*

The last question which we raised in the introduction was whether the second-best emissions (waste) tax in durable good economies under imperfect competition underinternalizes or overinternalizes marginal environmental (disposal) costs. Although the possibility of overinternalization cannot be excluded in the models considered in Chap. 3 and Sect. 4.1, we may yet conclude that under plausible conditions the second-best emissions (waste) tax rate falls short of marginal environmental (disposal) costs whenever the durable good producers possess commitment ability. The reason is that the second-best tax is greater than marginal damage only if either the equilibrium stock of the durable good or equilibrium emissions (solid waste) increases upon raising the emissions (waste) tax rate. However, empirical studies indicate that none of these conditions is satisfied so that underinternalization appears to be second-best optimal.

In contrast, if durable good producers do not possess commitment ability and if they sell their output, then overinternalization may be second-best optimal even without a positive correlation between equilibrium values and the environmental tax rate (Sect. 4.2). The reason for this result is again the presence of the producers' additional incentives for planned obsolescence and revenue raising. We have proved this overinternlization case by the way of example. Whether markets really satisfy the conditions for which the optimal tax exceeds marginal damage is ultimately an empirical issue.

Besides analyzing environmental aspects of durable goods, the models in the present study have also reexamined the validity of Swan's independence result. The main conclusion we can draw from our analysis is that the independence of product durability and market structure is sensitive to the combination of durable good taxation and the properties of the production cost function (Propositions 3.11 and 4.7).

It should be noted that our analysis abstracts from many features of durable good markets which might be important for environmental and resource policy of consumer durables. Let us finally discuss two examples which may be a starting point for future research. First, we did not take into account technical progress in the production of the durable good. Technical progress may take the form of learning effects. In the presence of learning effects, increasing product durability causes additional social costs which we did not identify in the present study: It reduces output so that the costs reduction due to the learning effects declines. Consequently, the efficient durability in the presence of learning effects is expected to be smaller than in the absence of learning effects. Second, consumption of durable goods is often characterized by fashion. The present study did not take into account this feature of durable good markets since we assumed throughout that a unit of the durable good provides consumers the same services regardless of its age. The presence of fashion in durable good consumption is also expected to reduce efficient product durability since consumers derive more utility from a new unit of the durable than from an old one. Other extensions of our models are, for example, making explicit the maintenance effort of the user of the durable good or the use intensity which, for simplicity, we have assumed to be fixed throughout. Finally, an important point for future research is to develop applied models (like general equilibrium models based, for example, on the models presented in Chap. 5) which are capable of deriving quantitative result and, thus, can be used for policy advise.

References

Abel, A. (1983), Market structure and the durability of goods, *Review of Economic Studies* **50**, 625–637.

Asch, P. and Seneca, J. (1976), Monopoly and external costs: An application of second-best theory to the automobile industry, *Journal of Environmental Economics and Management* **3**, 69–79.

Auernheimer, L. and Savings, P. (1977), Market organization and the durability of durable goods, *Econometrica* **45**, 219–228.

Ausubel, L. and Deneckere, R. (1987), One is almost enough for monopoly, *Rand Journal of Economics* **18**, 255–274.

Ausubel, L. and Deneckere, R. (1989), Reputation in bargaining and durable goods monopoly, *Econometrica* **57**, 511–531.

Baily, M., Hulten, C. and Campbell, D. (1992), Productivity dynamics in manufacturing plants, *Brookings Papers on Economic Activity: Microeconomics* pp. 187–249.

Bakke, V. (1974), A maximum principle for an optimal control problem with integral constraints, *Journal of Optimization Theory and Applications* **13**, 32–55.

Barnett, A. (1980), The Pigouvian tax rule under monopoly, *American Economic Review* **70**, 1037–1041.

Barrett, A. and Lawlor, J. (1997), Questioning the waste hierarchy: The case of a region with a low population density, *Journal of Environmental Planning and Management* **40**, 19–36.

Baumol, W. and Oates, W. (1988), *The Theory of Environmental Policy*, second edn, Cambridge: Cambridge University Press.

Bellmann, K. (1990), *Langlebige Gebrauchsgüter: Ökologische Optimierung der Nutzungsdauer*, Wiesbaden: DUV.

Beltratti, A., Chichilnisky, G. and Heal, G. (1997), Sustainable use of renewable resources, *in* G. Chichilnisky, G. Heal and A. Vercelli (eds), *Sustainability: Dynamics and Uncertainty*, Dordrecht: Kluwer Acadamic Publisher.

Blümel, W., von dem Hagen, O. and Pethig, R. (1986), The theory of public goods: A survey of recent issues, *Zeitschrift für die gesamte Staatswissenschaft* **142**, 241–309.

Brännlund, R. and Gren, I.-M. (1999), Green taxes in Sweden: A partial equilibrium analysis of the carbon tax and the tax on nitrogen in fertilizers, *in* R. Brännlund and I.-M. Gren (eds), *Green Taxes: Economic Theory and Empirical Evidence from Scandinavia*, Cheltenham: Edward Elgar.

Brisson, I. (1993), Packaging waste and the environment: Economics and policy, *Resources, Conservation, and Recycling* **8**, 183–292.

Buchanan, J. (1969), External diseconomies, corrective taxes and market structure, *American Economic Review* **59**, 174–177.

Bulow, J. (1986), An economic theory of planned obsolescence, *Quarterly Journal of Economics* **101**, 729–749.

Burnside, C. (1996), Production function regressions, returns to scale, and externalities, *Journal of Monetary Economics* **37**, 177–201.

Butz, D. (1990), Durable good monopoly and best-price provision, *American Economic Review* **80**, 1062–1076.

Calcott, K. and Walls, M. (1999), Policies to encourage recycling and 'design for environment': What to do when markets are missing? Discussion Paper, Victoria University of Wellington.

Calcott, K. and Walls, M. (2000), Can downstream waste disposal policies encourage upstream 'design for environment'?, *American Economic Review* **90**, 233–237.

Carle, D. and Blount, G. (1999), The suitability of aluminium as an alternative material for car bodies, *Materials and Design* **20**, 267–272.

Chichilnisky, G. (1996), Sustainable development: An axiomatic approach, *Social Choice and Welfare* **13**, 219–248.

Chichilnisky, G., Heal, G. and Vercelli, A. (1997), *Sustainability: Dynamics and Uncertainty*, Dordrecht: Kluwer Acadamic Publisher.

Choe, C. and Fraser, I. (1999), An economic analysis of household waste management, *Journal of Environmental Economics and Management* **38**, 234–246.

Choe, C. and Fraser, I. (2001), On the flexibility of optimal policies for green design, *Environmental and Resource Economics* **18**, 367–371.

Conn, W. (1977), Product life extension and design in the context of material balance, *in* D. Pearce and I. Walter (eds), *Resource Conservation*, New York: New York University Press.

Cremer, H. and Thisse, J.-F. (1999), On the taxation of polluting products in a differentiated industry, *European Economic Review* **43**, 575–594.

Dixit, A. (1986), Comparative statics for oligopoly, *International Economic Review* **27**, 107–122.

Ebert, U. (1992), Pigouvian tax and market structure: The case of oligopoly and different abatement technologies, *Finanzarchiv* **49**, 154–166.

Eichner, T. and Pethig, R. (1999), Product design and alternative market schemes for solid waste treatment and disposal. Volkswirtschaftliche Diskussionsbeiträge 73-99, University of Siegen.

Eichner, T. and Pethig, R. (2000), Recycling, producer responsibility and centralized waste management, *Finanzarchiv* **57**, 333–360.

Eichner, T. and Pethig, R. (2001), Product design and efficient management of recycling and waste treatment, *Journal of Environmental Economics and Management* **41**, 109–134.

Eichner, T. and Runkel, M. (2001), Efficient regulation of green design in a vintage durable good model. Discussion Paper, University of Siegen.

Eichner, T. and Runkel, M. (2003), Efficient management of product durability and recyclability under Utilitarian and Chichilnisky preferences, *Journal of Economics/Zeitschrift für Nationalökonomie* **80**, 43–75.

Feichtinger, G. and Hartl, R. (1986), *Optimale Kontrolle ökonomischer Prozesse*, Berlin: de Gruyter.

Fershtmann, C., Kamien, M. and Muller, E. (1992), Integral games: Theory and applications, *in* G. Feichtinger (ed.), *Dynamic Economic Models and Optimal Control*, Amsterdam: North-Holland.

Fishman, A. and Rob, R. (2000), Product innovation by a durable good monopoly, *Rand Journal of Economics* **31**, 237–252.

Fudenberg, D. and Tirole, J. (1991), *Game Theory*, Cambridge, London: MIT Press.

Fullerton, D. and Kinnaman, T. (1995), Garbage, recycling and illicit burning, *Journal of Environmental Economics and Management* **29**, 78–91.

Fullerton, D. and Kinnaman, T. (1996), Household responses to pricing garbage by the bag, *American Economic Review* **86**, 971–984.

Fullerton, D. and Wu, W. (1998), Policies for green design, *Journal of Environmental Economics and Management* **36**, 131–148.

Goering, G. (1992), Oligopolies and product durability, *International Journal of Industrial Organization* **10**, 55–63.

Goering, G. (1993), Learning by doing and product durability, *Economica* **60**, 311–326.

Goering, G. and Boyce, J. (1997), Optimal taxation of polluting durable goods monopolist, *Public Finance Review* **25**, 522–541.

Goering, G. and Boyce, J. (1999), Emissions taxation in durable goods oligopoly, *Journal of Industrial Economics* **47**, 125–143.

Golub, J. (ed.) (1998), *New Instruments for Environmental Policy in the EU*, London, New York: Routledge.

Hamilton, B. and Macauley, M. (1999), Heredity or environment: Why is automobile longevity increasing, *Journal of Industrial Economics* **47**, 251–261.

Hendel, I. and Lizzeri, A. (1999), Interfering with secondary markets, *Rand Journal of Economics* **30**, 1–21.

Huhtala, A. (1999), How much do money, inconvenience and pollution matter? Analysing households' demand for large-scale recycling and incineration, *Journal of Environmental Management* **55**, 27–38.

Jambor, A. and Beyer, M. (1997), Technical report: New cars – new materials, *Materials and Design* **18**, 203–209.

Jenkins, R. (1993), *The Economics of Solid Waste Reduction: The Impact of User Fees*, Hampshire, U.K.: Edward Elgar.

Kamien, M. and Muller, E. (1976), Optimal control with integral state equation, *Review of Economic Studies* **43**, 469–473.

Katsoulacos, Y. and Xepapadeas, A. (1995), Environmental policy under oligopoly with endogenous market structure, *Scandinavian Journal of Economics* **97**, 411–420.

Kleiman, E. and Ophir, T. (1966), The durability of durable goods, *Review of Economic Studies* **33**, 165–178.

Ko, I.-D., Lapan, H. and Sandler, T. (1992), Controlling stock externalities: Flexible versus inflexible Pigouvian corrections, *European Economic Review* **36**, 1263–1276.

Koopmans, T. (1960), Stationary ordinal utility and impatience, *Econometrica* **28**, 287–309.

Levhari, D. and Srinivasan, T. (1969), Durability of consumption goods: Competition versus monopoly, *American Economic Review* **59**, 102–107.

Lidgren, K. and Skogh, G. (1996), Extended producer responsibility: Recycling, liability, and guarantee funds, *The Geneva Papers on Risk and Insurance* **21**, 120–181.

Martin, D. (1962), Monopoly power and the durability of durable goods, *Southern Economic Journal* **28**, 271–277.

Michaelis, P. (1995), Product stewardship, waste minimization and economic efficiency: Lessons from Germany, *Journal of Environmental Planning and Management* **38**, 231–43.

Miranda, M. and Hale, B. (1997), Waste not, want not: The private and social costs of waste-to-energy production, *Energy Economics* **25**, 587–600.

Muller, E. and Peles, Y. (1990), Optimal dynamic durability, *Journal of Economic Dynamics and Control* **14**, 709–719.

OECD (1982), *Product Durability and Product Lifetime Extension: Their Contribution to Solid Waste Management*, Paris.

Palmer, P. and Walls, M. (1997), Optimal policies for solid waste disposal: Taxes, subsidies, and standards, *Journal of Public Economics* **65**, 193–205.

Palmer, P. and Walls, M. (1999), Extended product responsibility: An economic assessment of alternative policies. Discussion Paper 99-12, Resources for the Future, Washington.

Pigou, A. (1938), *The Economics of Welfare*, fourth edn, London: Macmillan & Co.

Pontryagin, L., Boltyanskii, V., Gamkrelidze, R. and Mishchenko, E. (1962), *The Mathematical Theory of Optimal Processes*, New York: Whiley-Interscience.

Ramm, W. (1974), On the durability of capital goods under imperfect market conditions, *American Economic Review* **64**, 787–796.

Requate, T. (1997), Green taxes in oligopoly if the number of firms is endogenous, *Finanzarchiv* **54**, 261–280.

Rousso, A. and Shah, P. (1994), Packaging taxes and recycling incentives: The German Green Dot Program, *National Tax Journal* **47**, 689–701.

Runkel, M. (1999), First-best and second-best regulation of solid waste under imperfect competition in a durable good industry. Volkswirtschaftliche Diskussionsbeiträge 81-99, University of Siegen.

Runkel, M. (2002), A note on emissions taxation in durable goods oligopoly, *Journal of Industrial Economics* **50**, 235–236.

Runkel, M. (2003a), Optimal emissions taxation under imperfect competition in a durable good industry, *Bulletin of Economic Research*. Forthcoming.

Runkel, M. (2003b), Product durability and extended producer responsibility in solid waste management, *Environmental and Resource Economics* **24**, 161–182.

Rust, J. (1986), When is it optimal to kill off the market for used durable goods, *Econometrica* **54**, 65–86.

Schmalensee, R. (1970), Regulation and the durability of goods, *Bell Journal of Economics and Management Science* **1**, 54–64.

Schmalensee, R. (1974), Market structure, durability and maintenance effort, *Review of Economic Studies* **41**, 277–287.

Schmalensee, R. (1979), Market structure, durability and quality: A selective survey, *Economic Inquiry* **17**, 177–196.

Seierstad, A. and Sydæster, K. (1987), *Optimal Control Theory with Economic Applications*, Amsterdam: North-Holland.

Shone, R. (1998), *Economic Dynamics*, Cambridge: Cambridge University Press.

Sieper, E. and Swan, P. (1973), Monopoly and competition in the market for durable goods, *Review of Economic Studies* **40**, 333–351.

Simpson, R. (1995), Optimal pollution taxation in a Cournot duopoly, *Environmental and Resource Economics* **6**, 359–369.

Soete, B. (1997), Ökoleasing – eine Innovation?, *Zeitschrift für angewandte Umweltforschung* **10**, 66–76.

Stahel, W. (1991), *Langlebigkeit und Materialrecycling: Strategien zur Vermeidung von Abfällen im Bereich der Produkte*, Essen: Vulkan Verlag.

Statistikbundesamt (2001), *Statistisches Jahrbuch 2001*, Wiesbaden: Metzler und Poeschel.

Staudt, E., Kunhenn, N., Scholl, M. and Interthal, J. (1997), *Die Verpackungsordnung: Auswirkungen eines umweltpolitischen Grossexperiments*, Innovation: Forschung und Management, Band 11, Bochum.

Su, T. (1975), Durability of consumption goods reconsidered, *American Economic Review* **65**, 148–157.

Swan, P. (1970), Durability of consumption goods, *American Economic Review* **60**, 884–894.

Tietenberg, T. (1996), *Environmental and Natural Resource Economics*, fourth edn, HarperCollins College Publishers.

Tinbergen, J. (1967), *Economic Policy: Principles and Design*, fourth edn, Amsterdam: North Holland.

Umweltbundesamt (1997), *Daten zur Umwelt*, Berlin: Schmidt-Verlag.

VDA (2001), Jahreszahlen zur Automobilproduktion. Webpage: http://www.vda.de/de/aktuell/statistik/jahreszahlen/index.html.

Waldman, M. (1993), A new perspective on planned obsolescence, *Quarterly Journal of Economics* **108**, 273–284.

Waldman, M. (1996), Planned obsolescence and the R&D decision, *Rand Journal of Economics* **27**, 583–595.

Weitzman, M. (1974), Prices vs. quantities, *Review of Economic Studies* **41**, 477–491.

Xepapadeas, A. (1997), *Advanced Principles in Environmental Policy*, Cheltenham: Edgar Elgar.

List of Figures

List of Tables

Printing and Binding: Strauss GmbH, Mörlenbach